Technology Trends in Wireless Communications

The Artech House Universal Personal Communications Series

Ramjee Prasad, Series Editor

Technology Trends in Wireless Communications

Ramjee Prasad
Marina Ruggieri

Artech House
Boston • London
www.artechhouse.com

Library of Congress Cataloging-in-Publication Data
Prasad, Ramjee
  Technology Trends in Wireless Communications/Ramjee Prasad, Marina Ruggieri.
      p. cm.—(Artech House universal personal communications series)
  Includes bibliographical references and index.
  ISBN 1-58053-352-3 (alk. paper)
  1. Wireless communication systems—Technological innovations.   I. Ruggieri, M.
  (Marina), 1961–   II. Title.

TK5103.2P7197 2003
621.382–dc21   2003041476

Library
University of Texas
at San Antonio

British Library Cataloguing in Publication Data
Prasad, Ramjee
  Technology trends in wireless communications.—
  (Artech House universal personal communications series)
  1. Wireless communication systems   2. Mobile communication systems
  3. Digital communications   4. Multimedia systems
  I. Title   II. Ruggieri, M. (Marina), 1961–
  621. 3'8456

  ISBN 1-58053-352-3

Cover design by Yekaterina Ratner

International Standard Book Number: 1-58053-352-3
Library of Congress Catalog Card Number: 2003041476

10 9 8 7 6 5 4 3 2 1

To our parents, Chandrakala and Sabita, Iole and Rino,
for their inspiration and motivation to meet our dreams.

Contents

Preface

दुःखेष्वनुद्विग्नमनाः सुखेषु विगतस्पृहः ।
वीतरागभयक्रोधः स्थितधीर्मुनिरुच्यते ॥ ५ ६॥

duḥkheṣv anudvigna-manāḥ
sukheṣu vigata-spṛhah
vīta-rāga-bhaya-krodhah
sthita-dhīr munir ucyate

One who is not disturbed in mind even amidst the threefold miseries or
elated when there is happiness, and who is free from attachment, fear and
anger, is called a sage of steady mind.

—The Baghavat Gita (2.56)

The use of mobile devices now surpasses that of traditional computers: wireless
users will hence soon be demanding the same rich multimedia services on their
mobile devices that they have on their desktop personal computers. In addition,
new services will be added, especially related with their mobile needs, such as
location-based information services.

High data rates will be necessary to carry multimedia communications,
and hence networks will be asked to deal with a multimedia traffic mix of data,
video, and voice packets, each having different transfer requirements. End users
will expect their link through the radio network to be interactive and robust and
will demand the wireless communicator be a small, low power, portable device.

In order to cope with all the highlighted challenges, new technical solutions at different levels have been already proposed and others are under
investigation.

Within the proposed frame, this book provides the reader the background and the hints necessary to look at the future technological trends of high-rate digital communications design.

The authors have been motivated to write this type of book for various reasons. Scientific research in the wireless multimedia communications field is growing fast: an update and harmonious compendium on the most recent technologies at both radio and network layer is needed. Furthermore, the design of wireless networks is indeed a multidisciplinary endeavour, where the interaction of different layers in the protocol stack has to be considered: this book has been conceived to cover several traditionally separated topics, thereby offering a complete guide to approach issues related to the wireless multimedia communication network planning (also thanks to the rich reference list in each chapter). In addition, future telecommunications systems supporting multimedia services have to provide users with access to services from both terrestrial and satellite fixed and mobile networks, according to given quality of service requirements, that might even change during the connection. Accounting for the limited availability of radio resources, the above requirement can be met through the development of flexible/adaptable radio interfaces. To this respect, the book gives the proper relevance to resource management strategies—such as power control, user admission techniques and congestion control—as well as to the adaptive transmission and reception techniques. They constitute a fundamental aspect for the provision of a variability degree of the quality of service and the effective exploitation of the limited radio resources, hence representing a hot research area. Finally, a future projection is certainly useful in this field, thus the book provides basic principles of future networks presently under investigation and an overview of the future application scenario.

The book is intended for use by graduate students approaching research activities in the wireless communications area and by professional engineers and project managers involved in wireless system design, aiming at consolidating their future vision of the wireless multimedia world.

The book is organized in nine chapters.

In Chapter 1 the main evolutionary steps from the current heterogeneous wireless networks towards the future integrated-services multimedia network are outlined, introducing the development of mobile networks from second to third generation, the vision beyond third generation, and the major technical challenges.

Chapter 2 provides the reader with an updated vision on multiple access protocols, focusing on the wireless domain and multimedia traffic. All those protocols adopted in current wireless communication systems are an adapted version of traditional and well-known mechanisms designed for wired networks or for specific wireless scenarios and applications. Furthermore, in order to support traffic with variable bit rates, different quality of service requirements, and other concepts like priority, scheduling, and quality of service support should be taken into account for the design of multiple access protocols.

Chapter 3 deals with the IP network issues: the mobile IP architecture, which is proposed for supporting mobility in Internet, is presented, as well as guidelines behind mobility. Quality of service provision, security, and routing issues are also addressed, first by introducing the main approaches. Protocols and mechanisms to fulfill security needs of new multimedia users in IP networks are also addressed in the chapter, together with proposals to solve some of the open issues in mobile IP.

Chapter 4 presents some proposed solutions and the ongoing research efforts to face the key issues of TCP/IP over wireless links for future wireless networks. An overview on the main mechanisms within TCP, which play an important role in using TCP over wireless links, is provided, together with the main configuration options that can be found in all modern versions of TCP. Possible approaches to split the connection in order to alleviate the effects of non-congestion-related packet losses are also described in the chapter, together with basic ARQ mechanisms and their interaction with the TCP protocol and improved link layer mechanisms that exploit some knowledge about the channel.

Chapter 5 provides basic concepts on channel adaptivity, specifically focusing on adaptive modulation and adaptive error control mechanisms, after recalling some results from the information theory. Furthermore, the Chapter 5 also addresses the trends in the implementation and design of adaptive transceivers.

Chapter 6 focuses on the main radio resource management functions and their implementation in different wireless networks, describing the most important radio resource management functions in GPRS, UMTS, and future wireless systems.

Chapter 7 is dedicated to real-time services. In particular, the main differences in the design of packet networks for real-time services as compared with the current design for non-real-time services are highlighted. Standards, protocols, and technologies needed to support new video applications over the Internet are introduced together with the main technologies to support voice over IP.

The main technical challenges of wireless personal area networks are presented in Chapter 8—after introducing the concept of personal area network—together with possible applications and devices as well as existing and emerging technologies for supporting these short-range communications.

In Chapter 9, moving from the hints and trends derived in the previous chapters, a future vision is presented to investigate a "Uniform Global Infrastructure." Of course, the future of communication systems appears quite unpredictable, and hence, it is rather difficult to provide a future scenario that can be agreed upon by all readers.

However, even if some readers have a different view for the future, we believe that what is proposed in this book can provide useful elements of discussion and a common basis for the development of the future multimedia world.

Acknowledgments

We would like to express our hearty appreciation to Dr. Ernestina Cianca (University of Rome "Tor Vergata"/Center for PersonKommunication, Aalborg University) for her invaluable contribution to this book. The contribution she gave has confirmed her deep commitment to scientific and technical matters, her professional capability, and, mostly importantly, her enthusiastic approach to solving complex problems. We would not have completed this book without her devoted support. In thanking her we would like to take this opportunity to wish her all the success in her academic career that she fully deserves.

We also wish to thank Ljupco Jorguseski of KPN Research Lab, the Netherlands, for supporting us in finishing Chapter 6. Finally, we appreciate the support of Junko Prasad in completing this book.

1

Introduction

The impressive evolution of mobile networks and the potential of wireless multimedia communications pose many questions to operators, manufacturers, and scientists working in the field. The future scenario is open to several alternatives: thoughts, proposals, and activities of the near future could provide the answer to the open points and dictate the future trends of the wireless world.

This book has been conceived as a tool—through its technical multilayer content and the vision elements—for those who may either wish to contribute to the definition and the development of the future scenario or just to be aware of it.

The focus of this book is on the future wireless multimedia communications, supporting all multimedia services, such as data, graphics, audio, images, and video, for different types of users: (1) users not physically wired to the network; (2) users able to access the network from many locations (i.e., nomadic users); and (3) users able to access the network while moving (i.e., mobile users).

In 2003–2005 the market of mobile multimedia services will experience a large increase, mainly driven by Internet-based data services [1–3]. The perspective of today's information society calls for a multiplicity of devices, including Internet Protocol (IP)-enabled home appliances, vehicles, personal computers, sensors, actuators, all of which are to be globally connected. Current mobile and wireless systems and architectural concepts must evolve in order to cope with these complex connectivity requirements. Scientific research in this truly multidisciplinary field is growing fast. New technologies, new architectural concepts, and new challenges are emerging [4–8]. A broader band knowledge, ranging over different layers of the protocol stack, is required by experts involved in research, design, and development aspects of future wireless networks.

Network design using the layered Open Systems Interconnection (OSI) architecture has been a satisfactory approach for wired networks especially as the communication links evolved to provide gigabit-per-second data rates and bit error rates (BERs) of 10^{-12}. Wireless channels typically have much lower data rates (on the order of a few Mbps), higher BERs (10^{-2} to 10^{-6}), and exhibit sporadic error bursts and intermittent connectivity. These performance characteristics change as network topology and user traffic also vary over time. Consequently, good end-to-end wireless network performance will not be possible without a truly optimized, integrated, and adaptive network design. Each level in the protocol stack should adapt to wireless link variations in an appropriate manner, taking into account the adaptive strategies at the other layers, in order to optimize network performance.

In this introductory chapter, the main steps of the evolution from the current heterogeneous wireless networks towards the future integrated-services multimedia network are outlined.

In Section 1.1, the development of mobile networks from second to third generation is considered; in Section 1.2 the vision beyond third generation is given, also pointing out the major technical challenges. Finally, an overview of the book is given in Section 1.3.

1.1   Evolution of Mobile Networks

The main achievements in the evolution of mobile networks, moving from second generation (2G) systems towards third generation (3G) through the so-called "evolved" 2G, are highlighted in what follows. The passage from generation to generation is not only characterized by an increase in the data rate, but also by the transition from pure circuit-switched (CS) systems to CS-voice/packed data and IP-core-based systems, as it is highlighted in Figure 1.1.

1.1.1   Evolved Second Generation Systems

Second generation systems represent a milestone in the mobile world, corresponding to the introduction of digital cellular communications. The evolution from the first generation (1G) of analog systems meant the passage to a new system, while maintaining the same offered service: voice.

The success of 2G systems, which extend the traditional Public Switched Telephone Network (PSTN) or Integrated Services Digital Network (ISDN) and allow for nationwide or even worldwide seamless roaming with the same mobile phone, has been enormous.

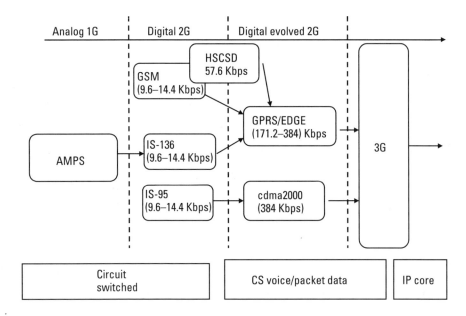

Figure 1.1 Evolution of cellular communications from 2G to 3G.

Today's most successful digital mobile cellular system is Global System for Mobile communications (GSM) [9–11] with users in more than 174 countries. In 2001 more than 600 million subscribers were reported, and projections give a number of subscriptions exceeding 1 billion by 2003. GSM is the only digital system in Europe, with over 320 million users.

In Japan the Personal Digital Cellular System (PDC) is operated. In the United States the digital market is divided into several systems, time division multiple access (TDMA)-based, code division multiple access (CDMA)-based, and GSM systems. This fragmentation has led to severe problems regarding coverage and service availability. About 32% of mobile subscribers in the United States and Canada still use the analog Advanced Mobile Phone Services (AMPS) system [12].

2G mobile systems are still mainly used for voice traffic. The basic versions typically implement a circuit-switched service, focused on voice, and only offer low data rates (9.6–14.4 Kbps).

Transitional data technologies between 2G and 3G have been proposed to achieve faster data rates sooner and at a lower cost than third generation systems. The evolved systems are characterized by higher data rates (64–384 Kbps) and packet data mode.

In what follows, some of the major evolved technologies of 2G systems are highlighted in order to provide to the reader the flavor of a key-step in the evolutionary path of mobile networks towards the multimedia era.

1.1.1.1   High-Speed Circuit-Switched Data

Within the frame of 2G technology, High-Speed Circuit-Switched Data (HSCSD) comes from the need to solve problems related to the slowness of GSM in data transmission.

In fact, GSM supports data transmissions with data rates up to 9.6 to 14.4 Kbps in circuit-switched mode and the transfer on signaling channels of small-size packets (up to 160 characters).

HSCSD was proposed by ETSI in early 1997. The key idea is to exploit more than one time slot in parallel among the eight time slots available with a proportional increment of the data rates [13, 14]. HSCSD allows the user to access, for instance, a company LAN, send and receive e-mail, and access the Internet whilst on the move. It is currently available to 90 millions subscribers across 25 countries.

On the other hand, HSCSD service does not effectively take advantage of the bursty nature of the traffic (e.g., Web browsing, e-mail, WAP). Channels are reserved during the connection. Furthermore, the exploitation of more time slots per user in a circuit-switched mode leads to a drastic reduction of channels available for voice users. For instance, four HSCSD users, each with four time slots assigned, prevent 16 voice users from accessing the network. Therefore, there is a need for packet-switched mode to provide a more efficient radio resource exploitation when *bursty* traffic sources are concerned.

HSCSD can be considered as a first step in the transitional technology between 2G and the packet-mode, higher rate evolved 2G systems (Figure 1.1).

1.1.1.2   i-mode

A great success in Japan has been obtained by the i-mode services, introduced in early 1999, which are provided by the packet-switched communication mode of the PDC system [15]. The i-mode hence represents a transitional step of PDC towards 3G.

The i-mode service utilizes compact HTML protocol, thus easing the interface to the Internet. Subscribers can send/receive e-mail and access a large variety of transactions, entertainment and database-related services, browsing Web sites and home pages. i-mode is very user-friendly and all instructions can be managed by only 10 keys.

1.1.1.3   General Packet Radio Service and Enhanced Data Rates for GSM Evolution

General Packet Radio Service (GPRS) and Enhanced Data Rates for GSM Evolution (EDGE) have been introduced as transitional data technologies for the evolution of GSM (Figure 1.1).

GPRS is the packet mode extension to GSM, supporting data applications and exploiting the already existing network infrastructure in order to save the operator's investments.

GPRS needs for a modest adaptation at radio interface level of GSM hardware. However, it adopts new physical channels and mapping into physical resources, as well as new radio resource management [16–18]. The new physical channel is called *52-multiframe* and it is composed of two 26 control multi-frames of voice-mode GSM.

High data rates can be provided since the GPRS users can exploit more than one time slot in parallel with the possibility, contrary to the HSCSD technology, to vary the number of time slot assigned to a user (e.g., to reduce them in case of scarcity of resources for the voice service). The maximum theoretical bit rate of the GPRS is 171.2 Kbps (using eight time slots). Current peak values are 20/30 Kbps. The 52-multiframe is logically divided into 12 *radio blocks* of four consecutive frames, where a radio block (20 ms) represents the minimum time resource assigned to a user. If the user is transmitting or receiving big flows of data, more than one radio block can be allocated to it. The whole set of these blocks received/transmitted by a mobile terminal during a reception/transmission phase forms the temporary block flow (TBF), which is maintained only for the duration of the data transfer. A session can consist of one or more TBFs that are activated during the transmission/reception phase. Each TBF is assigned a temporary flow identity (TFI) by the network, which is unique in both directions. For instance, during the reception, each mobile terminal listens to all the radio blocks flowing on the generic channel, but collects only the ones with the proper label (e.g., TFI). This mechanism simplifies the resource management in point-to-multipoint transmissions, like in the downlink (base station-mobile terminal), since each receiving station can pick up the proper blocks. Contrary to the GSM, GPRS service can flexibly handle asymmetric services by allocating a different number of time slot in uplink and downlink. Time slots can be allocated in two ways:

1. *On demand*, where the time slots not used by voice calls are allocated, and in case of resource scarcity for voice calls (congestion), time slots already assigned to GPRS service can be de-allocated;

2. *Static*, in which some time slots are allocated for GPRS and they cannot be exploited by voice calls.

In order to guarantee a minimum grade of service to GPRS users, the trade-off solution provides some channels statically allocated and the rest allocated on demand [18]. This allocation can be done dynamically with load supervision or capacity can alternatively be preallocated. Another new aspect of the GPRS with respect to GSM it is the possibility of specifying a quality of service (QoS) profile. This profile determines the service priority (high, normal, low), reliability and delay class of the transmission, and user data throughput [19, 20].

The radio link protocol provides a reliable link, while multiple access control (MAC) protocols control access with signaling procedures for radio channel and the mapping of link layer control (LLC) frames onto the GSM physical channels. Concerning the fixed backbone, the GPRS introduces two new networks elements: service GPRS support node (SGSN) and gateway GPRS support node (GGSN). In Figure 1.2, the GPRS architecture reference model is shown [21].

The SGSN represents for the *packet world* what the mobile switching center (MSC) represents for the *circuit world.* The SGSN performs mobility management [routing area update, attach/detach process, mobile station (MS) paging] as well as security tasks (e.g., ciphering of user data, authentication). GGSN tasks are comparable to the ones of a gateway MSC. It is not connected directly to the access network, but provides a means to connect SGSNs to other nodes or external packet data networks (PDNs). It also provides routing for packets coming from external networks to the SGSN where the MS is located as

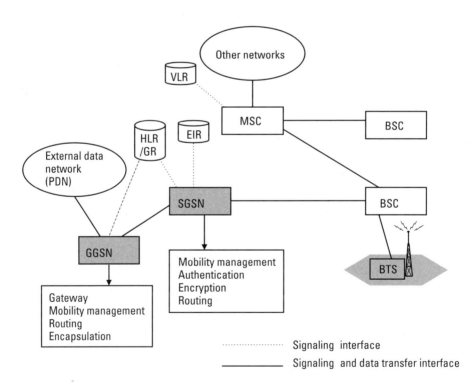

Figure 1.2 GSM-GPRS network architecture.

specified by the home location register (HLR). The new hardware boards for the BSC are called packet data units (PDUs) and their main functions are as follows: GPRS radio channels management (e.g., set-up/release); multiplexing of users among the available channels; power control, congestion control, broadcast of system information to the cells, and GPRS signaling from/to MS, base tranceiver station (BTS) and SGSN.

The evolution of the GPRS will be in the direction of improving the QoS by applying some of the concepts belonging to the 3G mobile systems (like the connection-oriented QoS) [19], powerful coding schemes, and more spectrally efficient modulation scheme, thus providing the user with services closer to real-time services [22].

In the evolutionary path to 3G systems, EDGE can be seen as a generic air interface for the efficient provision of higher bit rates, with respect to GSM, although by reusing the same GSM carrier bandwidth [23–27]. A typical GSM network operator deploying EDGE has a running GSM network, where EDGE can be introduced with minimal effort and costs.

EDGE uses enhanced modulation schemes, with respect to GSM, to increase the gross bit rate on the air interface and the spectral efficiency with moderate implementation complexity. Data rates up to 384 Kbps using the same 200-kHz wide carrier and the same frequencies as GSM (i.e., a data rate of 48 Kbps per time slot is available) can be achieved.

EDGE can be introduced incrementally, offering some channels that can switch between EDGE and GSM/GPRS. It will relieve both capacity and data rate bottlenecks in such a way that Internet and low bit rate audio-visual services become feasible on an on-demand basis. EDGE would also facilitate the "overnight delivery" of high-quality audio files [26].

1.1.2 Third Generation Systems

The evolution from 2G to 3G is characterized by a revolutionary change of focus from voice to mobile multimedia services, with the simultaneous support of several QoS classes in a single radio interface.

Third generation systems can provide higher data rates, thus enabling a much broader range of services [28–33]. The following types of services have been identified:

- Basic and enhanced voice services including applications such as audio conferencing and voice mail;
- Low data rate services supporting messaging, e-mail, facsimile;
- Medium data rate services for file transfer and Internet access at rates on the order of 64 to144 Kbps;

- High data rate services to support high-speed packet and circuit-based network access, and to support high-quality video conferencing at rates higher than 64 Kbps;

- Multimedia services, which provide concurrent video, audio, and data services to support advanced interactive applications;

- Multimedia services, also capable of supporting different quality of service requirements for different applications.

In 1985 the International Telecommunication Union (ITU) defined the vision for a 3G cellular system, at first called Future Public Land Mobile Telecommunications System (FPLMTS) and later renamed International Mobile Telecommunications-2000 (IMT-2000). ITU has two major objectives for the 3G wireless system: global roaming and multimedia services. The World Administrative Radio Conference (WARC'92) identified 1,885 to 2,025 and 2,110 to 2,200 MHz as the frequency bands that should be available worldwide for the new IMT-2000 systems. These bands will be allocated in different ways in different regions and countries [28]. A common spectral allocation along with a common air interface and roaming protocol design throughout the world can accomplish the global roaming capability. To support simultaneously new multimedia services that require much higher data rates and better QoS than only-voice services, the 3G wireless system envisages:

- Higher data rate services: up to 384 Kbps for mobile users, and 2 Mbps for fixed users, increasing to 20 Mbps;

- Increased spectral efficiency and capacity;

- Flexible air interfaces as well as more flexible resource management.

Compatibility with 2G systems is also one of the main goals of 3G systems. Different initiatives tried to unify the different proposals submitted to ITU in 1998 from ETSI for Europe, Association of Radio Industries and Broadcasting (ARIB) and Telecommunications Technology Council (TTC) for Japan, and American National Standard Institute (ANSI) for the United States.

1.1.2.1 Universal Mobile Telecommunications System

Universal Mobile Telecommunications System (UMTS) is the European version of IMT-2000. The UMTS Terrestrial Radio Access (UTRA) was approved by ITU in May 2000 [21]. A typical chip rate of 3.84 Mcps is used for the 5-MHz band allocation [34]. Wideband CDMA (W-CDMA) [28–31, 34–36], supported by groups in Japan (ARIB) and Europe, and backward-compatible with GSM, has been selected for the UTRA frequency division duplex (FDD) while TD-CDMA has been selected for the UTRA time division duplex (TDD). The introduction of TDD mode is mainly because of the asymmetric frequency bands designed by ITU. Also, the asymmetric nature of the data traffic on the forward and reverse links anticipated in the next generation wireless systems (e.g., Internet applications) suggests that TDD mode might be preferred over FDD.

Channelization codes are required to distinguish channels in both directions. Orthogonal variable spreading factor (OVSF) codes [36] are supported to provide variable data rates. User separation on the reverse link is achieved by user-specific either short or long scrambling codes. Short codes are very large Kasami codes of length 256, while long codes are Gold sequences of length 2^{46} [37]. Cell separation on the forward link is achieved by Gold sequences. The use of two access methods (FDD and TDD) together with the exploitation of variable bit rate techniques are important aspects in order to fulfill the flexibility requirement.

W-CDMA supports asynchronous mode operation where reception and transmission timings of different cell sites are not synchronized. It also supports forward and reverse fast closed loop power control [38] with an update rate of 1,600 Hz that is double with respect to the update rate in IS-95b.

In TDD mode, code, frequency, and time slot define a physical channel. In FDD mode, a physical channel is defined by its code and frequency and possibly by the relative phase. They have the following structure: a frame length of 10 ms organized in 15 time slots. The frame is the minimum transmission element in which the information rate is kept constant. The source bit rate can be different frame by frame, while the chip rate is always kept constant. A time slot has a duration of 10/15 ms and it is the minimum transmission element in which the transmission power is kept constant. Power control can update the transmission power level each time slot. Dual channel quadrature phase shift keying modulation is adopted on the reverse link, where the reverse link dedicated physical data channel (DPDCH) and the dedicated physical control channel are mapped to the I and Q channels, respectively. The I and Q channels are then spread to the chip rate with two different channelization codes and subsequently complex-scrambled by a mobile station-specific complex code. For multicode transmission, each additional reverse link DPDCH may be transmitted on the I or Q channel. Either short or long scrambling codes should be used on the reverse link.

Figure 1.3 provides an overview of the radio protocols architecture of the UMTS Radio Access Network (UTRAN).

Radio protocols can be divided in three levels: physical layer, data link layer, network layer. The link layer is divided in two sublayers: MAC and radio link control (RLC). MAC protocols provide an optimized radio access for packet data transmission through the statistical multiplexing of some users on a

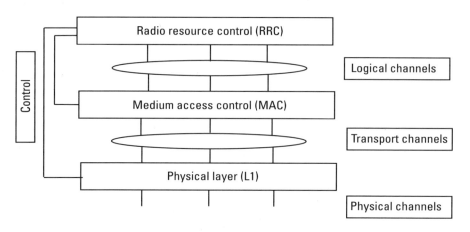

Figure 1.3 W-CDMA radio interface protocol architecture.

set of shared channels. They have a crucial importance in providing an effectiveness exploitation of the limited radio resources. The RLC provides a reliable transport of the information through retransmission error recovery mechanisms. The radio resource control (RRC) is part of the network layer and it is responsible for resource management. A *radio bearer* is allocated by the RRC with bit rates and QoS such that the service required by upper layers can be provided with the available resources at that moment. Note that resource management is an important issue in each mobile and wireless system, but it has a crucial importance in CDMA-based systems and quite different aspects with respect to the resource management in FDMA/TDMA systems like GSM. In this context, power control and radio admission control are key resource management mechanisms [39]. W-CDMA also has built-in support for future capacity enhancements as adaptive antennas, advanced receiver structures, and downlink transmit diversity. Furthermore, to improve the spectral efficiency to the extent possible, turbo codes [40, 41], capable of near Shannon limit power efficiency, have recently been adopted by the standards setting organizations in the United States, Europe, and Asia. For the UTRA/W-CDMA, the same constituent code is used for the rate 1/3-turbo code. Other codes are obtained by the "rate matching" process, where coded bits are punctured or repeated accordingly [42].

The possibility of handling simultaneously different kinds of services is supported by: variable bit rates techniques, multiplexing of several logical channels in the same dedicated transport channel, multiplexing of several transport dedicated channels in the same physical channel, common channel in uplink, and downlink channel shared by different users suitable for Internet applications. Furthermore, asynchronous transfer mode (ATM) has been chosen as transport technique in order to have a mechanism adaptable to different combination of multimedia traffic.

Figure 1.4 shows the architecture of a UMTS network. It consists of a core network (CN) connected with interface I_u to the UTRAN, which collects all the traffic coming from the radio stations. The UTRAN consists of a set of radio network subsystems (RNSs) connected to the CN through the I_u interface. Each RNS is responsible for the resources of its set of cells, and each node has one or more cells. A RNS is analogous to the BSS in the GSM-GPRS architecture and consists of a radio network controller (RNC) (which is analogous to the BSC) and one or more nodes B. The installation of a node B requires a complete replacement of the analogous BTS since it must handle the different air interface introduced in W-CDMA. A node B is connected to the RNC through the I_{ub} interface. An RNC separates the circuit-switched traffic (voice and circuit-switched data) from the packet-switched traffic and routes the former to the 3G-MSC and the latter to the 3G-SGSN. The 3G-MSC requires modifying the GPRS-MSC to handle new voice compression and coding algorithms and it processes the circuit-switched traffic routed to it by the RNC. The MSC then sends the data to a PSTN or another Public Land Mobile Network (PLMN). The packet-switched information is routed using the IP-over-ATM protocol specified by the ATM Adaptation Layer 5 (AAL5). The SGSN is modified to handle AAL5 traffic but performs the same function as in GPRS. Signaling and control functions between the mobile MS and the RAN typically depend on the radio technology whereas signaling and control functions between the MS and the CN are independent from the radio technology (i.e., access technique).

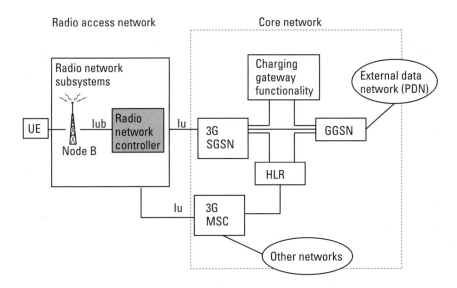

Figure 1.4 UMTS network architecture.

1.1.2.2 cdma1x

The cdma1x system is the evolution of the 2G IS-95 CDMA system and is supported by organization in the United States and Korea [43, 44].

In order to be backward-compatible with IS-95 networks, cdma1x retains many of the attributes of the IS-95 air interface design. Many carriers that deployed cdmaOne will choose cdma1x because the upgrade is quicker and less costly than UMTS. For example, in Korea, carriers have already rolled out IS-95C at 144 Kbps.

Adopting a synchronous mode of operation, transmission and reception timings of cell sites are synchronized by a single common timing source such as the Global Positioning System (GPS). Variable length Walsh codes (from 4 to 1,024 bits) are used for the spreading on supplemental channels to support various information rates [45]. In the forward link, Walsh codes are unique within channels of the same user as well as across different users in the same cell. Cell separation is performed by two PN sequences of length $2^{15} - 1$ chips, one for the in-phase channel and one for the in-quadrature channel. Shifted versions (in multiples of 64 chips) of the same sequences are used in different cells. Each cell uses a unique PN offset to distinguish its transmission from its neighboring cells. On the reverse link, cdma1x also uses Walsh codes to differentiate between channels from the same user. User separation is achieved by user-specific long PN codes. All the users in all cells use the same long code. The transmission from different users, however, is offset by a different number of bits. This offset is achieved by using the electronic serial number (ESN), which is unique to each mobile station. cdma1x introduces dedicated and common control channels to provide efficient packet data services. Fast power control with an update rate of 800 Hz is applied only in the reverse link of IS-95. Its implementation in the forward link of cdma1x provides significant performance improvements in low mobility environment where most of the high data rate applications will occur [40–48].

Both IS-95 and cdma1x employ a common pilot channel shared by all mobiles to provide a reference signal to receivers thus helping the coherent demodulation. In cdma1x this common pilot channel is a code-multiplexed channel using Walsh codes for orthogonal spreading. The incorporation of a pilot channel also on the reverse link offers significant performance gain by providing a coherent phase reference for coherent demodulation at the BS. It also reduces the power control loop delay. When beam-forming is applied to cover a smaller portion of a cell, the receivers would require an additional dedicated pilot for reliable channel estimation. Channel estimations will not be accurate if the reference pilot traverses a different path compared to the data signal. Thus, dedicated and common auxiliary channels are introduced to take advantage of smart antennas.

Systems can provide IS-95B and cdma1x services simultaneously to the MS. A new *burst mode* capability is defined that addresses the technical issues that arise when direct sequence spread spectrum has to support higher-data rate packet services. Turbo codes are also adopted in the cdma1x standard [48]. cdma1x also includes a sophisticated MAC feature to support effectively very high data rate services (up to 2 Mbps) and multiple concurrent data and voice services. Enhancements over IS-95B include the introduction of the suspended and control hold states for packet data MAC. Additional enhanced features can be found in [49].

1.1.3 Wireless Local Area Networks

Another important aspect of the evolutionary path of wireless networks is represented by wireless local area networks (WLANs) [33]. The WLAN market is exploding, with reported yearly growth figures of 300% [9].

WLAN systems are a technology that can provide very high data rate applications and individual links (e.g., in company campus areas, conference centers, airports) and represents an attractive way of setting up computers networks in environments where cable installation is expensive or not feasible. They represent the coming together of two of the fastest-growing segments of the computer industry: LANs and mobile computing, thus recalling the attention of equipment manufactures.

This shows their high potential and justifies the big attention paid to WLAN by equipment manufacturers.

Whereas in the early beginning of WLANs several proprietary products existed, nowadays they are mostly conform to the Institute of Electrical and Electronics Engineering (IEEE) 802.11b (also known as Wi-Fi) standard [50]. It operates in the unlicenced 2.4-GHz band at 11 Mbps and it is currently extended to reach 20 Mbps [9]. A description of the MAC can be found in [51]. Figure 1.5 depicts different WLAN standards.

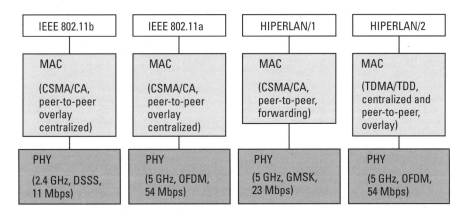

Figure 1.5 WLAN standards.

In Europe, another standard called High Performance Local Area Network (HIPERLAN) has made use of the unlicensed spectrum in the 5-GHz band where more bandwidth is available, for higher-rate (HIPERLAN/1: 20 Mbps), higher-quality multimedia systems [52]. Next generation wireless LAN standards, including IEEE 802.11a and HIPERLAN/2, offer higher performance and greater range of capabilities. They have harmonized physical layers, all OFDM with a maximum data rate of 54 Mbps [53]. A newly formed ETSI working group called Broadband Radio Access Network (BRAN) is working on extension to HIPERLAN standard. Parallel to HIPERLAN/2 standardization work, the Multimedia Mobile Access Communications (MMAC) Association in Japan started to develop different high-speed radio access system for business and home applications at 5 GHz. One of these systems for business in public and corporate networks is also aligned with HIPERLAN/2 at both the physical layer and data link layer. In this way, global roaming in these three regions is achieved.

The HIPERLAN/2 (54 Mbps) standard, ratified in 2000, differs from the IEEE 802.11a counterparts; in fact, HIPERLAN/2 systems use a connection-oriented protocol intended to support a variety of voice, data, and multimedia services. Furthermore, IEEE 802.11a/b is a wireless Ethernet with carrier sense multiple access/collision avoidance MAC. HIPERLAN/2 is centralized, using TDD/TDMA and thus allows for efficient quality of service over the radio interface.

1.1.4   Ad Hoc Networks and Wireless Personal Area Networks

Many WLANs of today need an infrastructure network that provides access to other networks and include MAC. Ad hoc wireless networks do not need any infrastructure. In these systems mobile stations may act as a relay station in a multihop transmission environment from distant mobiles to base stations. Mobile stations will have the ability to support base station functionality. The network organization will be based on interference measurements by all mobiles and base stations for automatic and dynamic network organization according to the actual interference and channel assignment situation for channel allocation of new connections and link optimization. These systems will play a complementary role to extend coverage for low power systems and for unlicensed applications. A central challenge in the design of ad hoc networks is the development of dynamic routing protocols that can efficiently find routes between two communication nodes. A Mobile Ad Hoc Networking (MANET) working group has been formed within the Internet Engineering Task Force (IETF) to develop a routing framework for IP-based protocols in ad hoc networks [54]. Another challenge is the design of proper MAC protocols for multihop ad hoc networks [55].

LANs without the need for an infrastructure and with a very limited coverage are being conceived for connecting different small devices in close proximity without expensive wiring and infrastructure. The area of interest could be the personal area about the person who is using the device. This new emerging architecture is indicated as wireless personal area network (WPAN). The concept of personal area network hence refers to a space of small coverage (less than 10m) around a person where ad hoc communication occurs [56], and is also referred as personal operating space (POS). The network is aimed at interconnecting portable and mobile computing devices such as laptops, personal digital assistants (PDAs), peripherals, cellular phones, digital cameras, headsets, and other electronics devices.

Bluetooth is an example of WPAN [57]. Some market-forecast analysis predicts that there will be some 1.4 billion Bluetooth devices in operation by the 2005 [58]. Therefore, Bluetooth is poised to play a large part in the future of personal wireless networking. The Bluetooth Special Interest Group has produced a specification [59] for wirelessly connecting information devices in a small, personal area. This specification represents the first step in establishing a new technology [60].

After introducing the WPAN concept, wireless connectivity can be seen in three basic categories, primarily based on coverage and mobility requirements [61]: wireless wide area network (WWAN) (i.e., the cellular systems), WLAN, and WPAN.

1.2 Vision

Economic and technical trends together with applications requirements will drive the future of mobile communications. Forecast of the mobile communications market is shown in Figure 1.6 for Japan, as an example [62].

The number of mobile communications subscribers is expected to reach 81 million by 2010 [62, 63]. From the current increasing ratio, this number will be saturated around 2006 with a penetration rate of approximately 70%. Although the number of subscribers will become saturated, traffic will still increase. In particular, Figure 1.6 shows the trend of Mobile Internet that represents the main driver for multimedia applications. The number of subscribers has increased much faster than expected and it will continue to grow through the 2000s. It is expected by the UMTS Forum that in Europe in 2010 more than 90 million mobile subscribers will use mobile multimedia services and will generate about 60% of the traffic in terms of transmitted bits. Additional frequency assignment will be necessary for 3G to accommodate the growing demand. The bandwidth to be added is assumed to be 160 MHz in 2010. However, the added bandwidth greatly depends on the growth ratio of traffic per

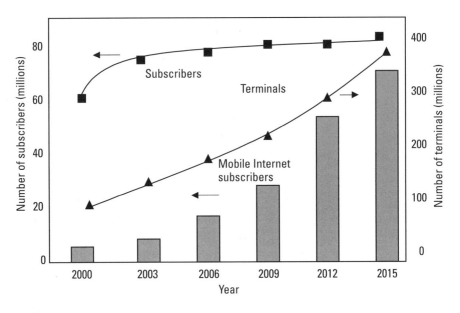

Figure 1.6 Mobile communications market forecast for Japan.

subscriber. Therefore, the study of high-capacity cellular systems with improved spectrum efficiency and new band is necessary to accommodate growing traffic in 2010 and beyond. Higher data rates and wireless Internet access are key components of the future mobile communications systems. But they are actually key concepts also in 3G systems. Data rates up to 8 Mbps will be possible without making any drastic change in the current standard. Future mobile communications systems should bring something more than only faster data or wireless Internet access [34, 62, 64]. Something that we are missing today (and even in the 3G) is the flexible interoperability of various existing networks like cellular, cordless, WLAN type systems, systems for short connectivity and wired systems. It will be a huge challenge to integrate the whole worldwide communication infrastructure to form one transparent network allowing various ways to connect into it depending on the user's needs as well as the available access methods. The heterogeneity of various accesses can be overcome either by using multimode terminals, additional network services, or by creating a completely new network system that will implement the envisioned integration. The first option implies only further development of the existing networks and services and cannot be very flexible. The second option is more profound and can result in a more efficient utilization of networks and available spectrum. The creation of this new network requires a completely new design approach. So far, most of the existing systems have been designed in isolation without taking into account a possible interworking with other access technologies. Their system design is mainly based on the traditional vertical approach to support a certain set of services with a particular technology. The UTRA concept has already combined the FDD and TDD components to support the different symmetrical and asymmetrical service needs in a spectrum-efficient way. This is the first step to a more horizontal approach [45, 65] where different access technologies will be combined into a common platform to complement each other in an optimum way for different service requirements and radio environments. Due to the dominant role of IP-based data traffic, these access systems will be connected to a common, flexible, and seamless IP-based core network. This results in a lower infrastructure cost, faster provisioning of new features, and easy integration of new network elements and could be supported by technologies like JAVA Virtual Machines and CORBA. The vision of the seamless future network is shown in Figure 1.7 [64].

The mobility management will be part of a new Media Access System as the interface between the core network and the particular access technology to connect a user via a single number for different access systems to the network. Global roaming for all access technologies is required. The internetworking between these different access systems in terms of horizontal and vertical handover and seamless services with service negotiation with respect to mobility,

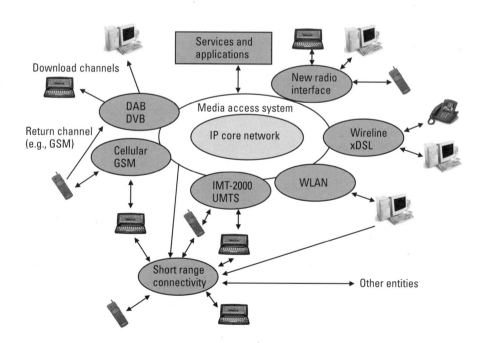

Figure 1.7 Seamless future network including a variety of internetworking access systems.

security, and QoS, will be a key requirement. The latter will be handled in the common Media Access System and the core network. Multimode terminals and new appliances are also key components to support these different access technologies of the common platform seamlessly from the user perspective. These terminals may be adaptive, based on high signal processing power. Therefore, the concept of software defined radio, supported by software downloading, could be a key technology in the future perspective.

To make the vision of systems beyond third generation happen, many technical challenges have to be solved by extensive research activities at different layers. In spite of the 2-Mbps data rates achievable by 3G systems, the overall economic capacity of these systems could still be only a small fraction of the actual need of the seamless information mobility. Radio remains a bottleneck for which no break-through solution exists today. The common aim is to exploit effectively and flexibly the limited radio resources [29, 30], taking into account the interaction among different layers of the protocol stack. The best efficiency is achieved if the system is able to adapt to environmental conditions that change in time, location, and even available service mixture. This leads to investigate adaptive radio interface and network systems where diversity means [66–75], coding method as well as type and gain [76–78], modulation method or order (M-QAM, M-PSK, multicode/single code, multicarrier/single carrier) [79–81], data rate, data packet size, channel allocation, and service selection can be adapted effectively. New radio access concepts [82], advanced reception techniques [83–88], smart antennas technology [69–78], and resource management issues [89–96] are the main research areas that could contribute to the achievement of the major goals of future generation wireless systems [97].

1.3 Preview of the Book

Chapter 2 is an overview of the mechanisms adopted in wireless systems for sharing common resources [i.e., multiple access protocols (MAP)]. First, traditional and well-known MAPs are recalled according to a classification in: random access protocols (e.g., ALOHA), contentionless protocols (frequency/time division multiple access, PRMA), and CDMA-based protocols. Current proposals for MAPs suitable for wireless multimedia traffic are presented and key criteria to be followed in the design of MAC for future communication scenarios are highlighted. In particular, MAC protocols that are being proposed for supporting real and non-real-time services in the standard WCDMA and UTRA TDD are described.

Chapter 3 deals with network issues in IP networks such as mobility support in IP networks, routing algorithms, security issues (IPSec, AAA), and provision of quality of service for real-time services in IP networks. Therefore,

Integrated Services (IntServ) and Differentiated Services (DiffServ) architectures are overviewed. Moreover, the Multi-Protocol Label Switching (MPLS) approach is introduced as a promising technology for delivering QoS on IP-based networks. The main approaches generally proposed for IP fixed networks are first discussed and then some extensions for mobile networks are given, focusing on Mobile IP. Open research issues for Mobile IP version 4 and Mobile IP version 6 are highlighted.

TCP/IP protocols are tuned to perform well in wired networks where the main cause of packet loss is network congestion. In wireless links, where the packet losses can be due to the lossy channel, they show performance degradation. This topic has been studied extensively in the literature. In Chapter 4, main approaches to this problem are described. The approach proposed recently by relevant literature in attempting to solve this problem presents two alternatives: (1) to achieve a more reliable channel in an attempt to "hide" the wireless link from the upper protocol layers (ARQ, FEC); and (2) to make the TCP sender aware of the existence of wireless hops and realize that some packet losses are not due to congestion (indirect TCP, snooping TCP, selective retransmissions). The chapter also describes the new trends of the research that shows a particular attention to the interaction among physical layer and upper layers.

Chapter 5 provides basic concepts on channel adaptivity. The focus is on adaptive modulation and adaptive error control mechanisms that adjust the transmission related parameters, like modulation level, symbol transmission rate, or coding rate, according to the instantaneous fading channel conditions for the higher radio spectral efficiency is provided. Some results from information theory are recalled to show the limitations of these techniques and motivate further research on the practical and design issues that have to be faced to reach performance close to theoretical limits.

Finally, trends in the implementation and design of adaptive transceivers are discussed, highlighting the need and the meaning of a cross-layered approach. The dominant trend is the implementation of radio systems as much as possible through digital processing, by developing high-performance signal processing processors.

Resource management strategies (power control, admission control, congestion control) play a key role in providing variable QoS. In Chapter 6, the main solutions envisaged for 3G are described and new concepts are introduced.

In Chapter 7, circuit-switched networks and packet-switched networks are compared for carrying real-time services. Then, the major standards and signal processing methods necessary for video streaming systems in wireless networks (MPEG family) are described and the joint source and channel coding for a wireless channel is carried out. Issues and technologies for delivering voice over IP are presented.

In Chapter 8, the concept and architectures of PANs are described. The current networks available for this type of service are discussed. Two main types of personal area networks can be distinguished: (1) a network carried on the body and thus a nondynamic network; and (2) a network among different moving entities (between people, between person and printer upon entering room, etc.). In the last case, ad hoc networks are important. The chapter discusses the main technical challenges for the development of wireless PANs such as power efficiency, service discovery/selection, security, ad hoc networking coexistence, and interference-reduction techniques.

In the last chapter, authors draw a scenario for future wireless multimedia communications, and a scientific approach to be followed towards the realization of this scenario is proposed. Techniques and technologies previously introduced in the book represent the needed background to understand and to envisage these future trends. Moreover, research areas of interest in the wireless communication world are highlighted.

References

[1] Prasad, R., M. Ruggieri, "Special Issue on the Future Strategy for the New Millenium Wireless World," *Wireless Personal Communications*, Vol. 17, Nos. 2–3, June 2001.

[2] "Technology Forecast: 2001–2003-Mobile Internet: Unleashing the Power of Wireless," PriceWaterHouseCoopers Technology Center.

[3] Arroyo-Fernandez, B., et al., "Life After Third-Generation Mobile Communications," *IEEE Communications Magazine*, Vol. 39, No. 8, Aug. 2001, pp. 41–42.

[4] Dinis, M., and J. Fernandes, "Provision of Sufficient Transmission Capacity for Broadband Mobile Multimedia: A Step Toward 4G," *IEEE Communications Magazine*, Vol. 39, No. 8, Aug. 2001, pp. 46–54.

[5] Tjelta, T., et al., "Future Broadband Radio Access Systems for Integrated Services with Flexible Resource Management," *IEEE Communications Magazine*, Vol. 39, No. 8, Aug. 2001, pp. 56–63.

[6] Robles, T., et al., "QoS Support for an All-IP System Beyond 3G," *IEEE Communications Magazine*, Vol. 39, No. 8, Aug. 2001, pp. 64–72.

[7] Becchetti, L., P. Mahonen, and L. Munoz, "Enhancing IP Service Provision over Heterogeneous Wireless Networks: A Path Toward 4G," *IEEE Communications Magazine*, Vol. 39, No. 8, Aug. 2001, pp. 74–81.

[8] Mehta, M., et al., "Reconfigurable Terminals: An Overview of Architectural Solutions," *IEEE Communications Magazine*, Vol. 39, No. 8, Aug. 2001, pp. 82–89.

[9] Pereira, J. M., "Balancing Public and Private in Fourth Generation," *Proceed. IEEE PIMRC 2001*, San Diego, Sept./Oct. 2001, pp. 125–132.

[10] ETSI (1991b), *General Description of a GSM PLMN*, European Telecommunications Standards Institute, GSM recommendations 01.02.

[11] Lin, Y.-B., and I. Chlamtac, *Mobile Network Protocols and Services*, New York: Wiley, 2000.

[12] Krenik, W. R.. "Wireless User Perspectives in the United States," *Wireless Personal Communications Journal*, Kluwer, Vol. 22, No. 2, Aug. 2002, pp. 153–160.

[13] Schiller, J., *Mobile Communications*, Addison-Wesley, 2000.

[14] Prasad, N. R., "GSM Evolution Towards Third Generations UMTS/IMT-2000," *Proc. IEEE International Conference on Personal Wireless Communication*, 1999, pp. 50–54.

[15] Nakajima, N., "Future Communications Systems in Japan," *Wireless Personal Communications*, Vol. 17, No. 2, June 2001, pp. 209–223.

[16] Digital cellular telecommunications system (Phase 2+); General Packet Radio Service (GPRS); Overall description of the GPRS radio interface; stage 2. GSM 03.64, version 7.0.0, release 1998.

[17] Cai, J., and D. J. Goodman, "General Packet Radio Service in GSM," *IEEE Communications Magazine*, Oct. 1997.

[18] Lin, P., and Y. Lin, "Channel Allocation for GPRS," *IEEE Transaction on Vehicular Technology*, Vol. 50, No. 2, March 2001, pp. 375–387.

[19] Priggouris, G., S. Hdjiefthymiades, and L. Merakos, "Supporting IP QoS in the General Packet Radio Service," *IEEE Network*, Sept.–Oct. 2000, pp. 8–17.

[20] Sarikaya, B., "Packet Mode in Wireless Networks: Overview of Transition to Third Generation," *IEEE Communication Magazine*, Vol. 38, No, 9, Sept. 2000, pp. 164–172.

[21] Muratore, F., *UMTS: Mobile Communications for Future*, New York: John Wiley & Sons, 2000.

[22] Lin, Y. B., A. Pang, and M. F. Chang, "VGPRS a Mechanism for Voice over GPRS," *Proc. International Conference on Distributed Computing Systems Workshop*, 2001, pp. 435–440.

[23] Third Generation Partnership Project (3GPP), UMTS and GSM Standards Data Base, http://www.3gpp.org.

[24] van Nobelen, R., et al., "An Adaptive Radio Link Protocol with Enhanced Data Rates for GSM Evolution," *IEEE Personal Communications*, Vol. 6, No. 1, Feb. 1999, pp. 54–64.

[25] Furuskar, A., et al., "EDGE: Enhanced Data Rate for GSM and TDMA/136 Evolution," *IEEE Personal Communications*, Vol. 6, No. 3, June 1999, pp. 56–66.

[26] Berthet, A., R. Visoz, and P. Tortelier, "Sub-Optimal Turbo-Detection for Coded 8-PSK Signals over ISI Channels with Application to EDGE Advanced Mobile System," *Proc. 11th IEEE International Symposium on Personal, Indoor and Mobile Radio Communications*, PIMRC 2000, Vol. 1, 2000, pp. 151–157.

[27] Gerstacker, W. H., and R. Schober, "Equalisation for EDGE Mobile Communications," *Electronics Letters*, Vol. 36, No. 2, Jan. 2000, pp. 189–191.

[28] Prasad, R., *Wideband CDMA for Third-Generation Mobile Systems*, Norwood, MA: Artech House, 1998.

[29] Prasad, R., W. Mohr, and W. Konhauser, *Third-Generation Mobile Communication System*, Norwood, MA: Artech House, 2000.

[30] Ojanpera, T., and R. Prasad, *WCDMA: Towards IP Mobility and Mobile Internet*, Norwood, MA: Artech House, 2001.

[31] Prasad, R., *Towards a Global 3G System: Advanced Mobile Communications in Europe*, Norwood, MA: Artech House, 2001.

[32] Schiller, J., *Mobile Communications*, Reading, MA: Addison-Wesley, 2000.

[33] Prasad, N. R., and A. Prasad, *WLAN Systems and Wireless IP for Next Generation Communication*, Norwood, MA: Artech House, 2002.

[34] Rapeli, J., "Future Directions for Mobile Communications Business, Technology and Research," *Wireless Personal Communications*, Vol. 17, No. 2, June 2001, pp. 155–173.

[35] Zeng, M., A. Annamalai, and V. Barghava, "Recent Advances in Cellular Wireless Communications," *IEEE Communications Magazine*, Sept. 1999, Vol. 37, No. 9, pp. 128–138.

[36] Holma, H., and A. Toskala, *WCDMA for UMTS*, New York: John Wiley & Sons, 2000.

[37] Dinan, E. H., and B. Jabbari, "Spreading Codes for DS-CDMA and Wideband CDMA Cellular Networks," *IEEE Communications Magazine*, Sept. 1998, pp. 48–54.

[38] Kauffmann, P., "Fast Power Control for Third Generation DS-CDMA Mobile Radio System," *Proc. 2000 International Zurich Seminar on Broadband Communications*, 2000, pp. 9–13.

[39] Jorguseski, L., J. Farserotu, and R. Prasad, "Radio Resource Allocation in Third Generation Mobile Communication Systems," *IEEE Communication Magazine*, Vol. 39, No. 2, Feb. 2001, pp. 117–123.

[40] Benedetto, S., and G. Montorsi, "Unveiling Turbo Codes: Some Results on Parallel Concatenated Coding Schemes," *IEEE Transactions on Information Theory*, Vol. 42, No. 2, 1996, pp. 409–428.

[41] Perez, L. C., J. Seghers, D. J. Costello, Jr., "A Distance Spectrum Interpretation of Turbo Codes," *IEEE Transactions on Information Theory*, Vol. 42, 1996, pp. 1698–1709.

[42] 3GPP, Technical Specification Group, Radio Access Network, Working Group 1, Multiplexing and Channel Coding (FDD).

[43] Knisely, D. N., et al., "Evolution of Wireless Data Services: IS-95 to cdma2000," *IEEE Communications Magazine*, Vol. 36, No. 10, Oct. 1998, pp. 140–149.

[44] Rao, Y. S., and A. Kripalani, "cdma2000 Mobile Radio Access for IMT 2000," IEEE International Conference on Personal Wireless Communication, 1999, pp. 6–15.

[45] Lee, D., H. Lee, and L. B. Milstein, "Direct Sequence Spread Spectrum Walsh-QPSK Modulation," *IEEE Transaction on Communications*, Vol. 46, No. 9, Sept. 1998, pp. 1227–1232.

[46] Chulajata, T., and H. M. Kwon, "Combinations of Power Controls for cdma2000 Wireless Communications System," *Proc. 52nd Vehicular Technology Conference, 2000, IEEE VTS Fall VTC 2000*, pp. 638–645.

[47] Chih-Lin, I., S. Nanda, "Load and Interference Based Demand Assignment for Wireless CDMA Netwroks," *Proc. IEEE Globecom*, 1996, pp. 235–241.

[48] "Standards for cdma2000 Spread Spectrum Systems," EIA/TIA IS-2000.1-6.

[49] Etemad, K., "Enhanced Random Access and Reservation Scheme in CDMA2000," *IEEE Personal Communications*, Apr. 2001, pp. 30–36.

[50] IEEE 802.11, IEEE Standard for Wireless LAN Medium Access Control (MAC) and Physical Layer (PHY) Specifications, Nov. 1997.

[51] Crow, B. P., et al., "IEEE 802.11 Wireless Local Area Network," *IEEE Communications Magazine*, Sept. 1997, pp. 116–126.

[52] ETSI, "Radio Equipment and Systems, High Performance Radio Local Area Network (HIPERLAN) Type 1," European Telecommunication Standard, ETS, 300–652, Oct. 1996.

[53] Van Nee, R., and R. Prasad, *OFDM for Wireless Multimedia Communications*, Norwood, MA: Artech House, 2000.

[54] Macker, J. P., and M. S. Corson, "Mobile ad hoc Networking and the IETF," *Proc. ACM, Mobile Computing and Communications*, Vol. 2, No. 1, Jan. 1998.

[55] Xu, S., and T. Saadawi, "Does the IEEE 802.11 MAC Protocol Work Well in MultiHop Wireless Ad Hoc Networks?" *IEEE Communications Magazine*, Vol. 39, No. 6, June 2001, pp. 130–137.

[56] Niemegeers, I. G., and S. M. H. De Groot, "From Personal Area Networks to Personal Networks: A User Oriented Approach," *Wireless Personal Communications*, Vol. 22, No. 2, Aug. 2002, pp. 175–186.

[57] Bisdikian, C., "An Overview of the Bluetooth Wireless Technology," *IEEE Communications Magazine*, Vol. 39, No. 12 , Dec. 2001, pp. 86–94.

[58] "Bluetooth 2000: To Enable the Star Trek Generation", Cahners In-Stat Group, MM00-09BW, June 2000.

[59] Bluetooth Special Interest Group, "Specification of the Bluetooth System," Dec. 1999.

[60] Siep, T. M., et al., "Paving the Way for Personal Area Network Standards: An Overview of the IEEE P802.15 Working Group for Wireless Personal Area Networks," *IEEE Personal Communications*, Vol. 7, No.1, Feb. 2000, pp. 37–43.

[61] Jha, U., "Wireless Landscape—Need for Seamless Connectivity," *Wireless Personal Communications*, Vol. 22, No. 2, Aug. 2002, pp. 275–283.

[62] Ohmori, S., Y. Yamao, and N. Nakajima, "The Future Generations of Mobile Communications Based on Broadband Access Methods," *Wireless Personal Communications*, Vol. 17, No. 2, June 2001, pp. 175–190.

[63] Japanese Telecommunications Technology Council, "A Partial Report on Technical Conditions on Next-Generation Mobile Communications System," Sept. 1999.

[64] Mohr, W., "Development of Mobile Communications Systems Beyond Third Generation," *Wireless Personal Communications*, Vol. 17, No. 2, June 2001, pp. 191–207.

[65] Chaudhury, P., W. Mohr, and S. Onoe, "The 3GPP Proposal for IMT-2000," *IEEE Communications Magazine*, Vol. 37, No. 12, 1999, pp. 72–81.

[66]   Poor, H. V., and G. W. Wornell, *Wireless Communications-Signal Processing Perspective*, Englewood Cliffs, NJ: Prentice Hall, 1998.

[67]   Andersen, J. B., "Array Gain and Capacity for Known Random Channels with Multiple Element Arrays at Both Ends," *IEEE Journal on Selected Areas in Communications*, Vol. 18, No. 11, Nov. 2000, pp. 2172–2178.

[68]   Foschini, G. J., and M. J. Gans, "Capacity When Using Multiple Antennas at Transmit and Receive Sites and Rayleigh-Faded Matrix Channel Is Unknown to the Transmitter," in *Advanced in Wireless Communications*, J. M. Holtzmann and M. Zorzi, (eds.), Kluwer Academic Publishers, 1998.

[69]   *IEEE Personal Communications*, Vol. 5, Feb. 1998.

[70]   Pattan, B., *Robust Modulation Methods and Smart Antennas in Wireless Communications*, Englewood Cliffs, NJ: Prentice Hall, 2000.

[71]   Wolniansky, P. V., et al., "V-BLAST: An Architecture for Realizing Very High Data Rates over the Rich-Scattering Wireless Channel," *Proc. ISSSE-98*, Pisa, Italy, Sept. 29, 1998.

[72]   Foschini, G. J., "Layered Space-Time Architecture for Wireless Communication in a Fading Environment when Using Multi-Element Antennas," *Bell Labs Technical Journal*, 1996, pp. 41–59.

[73]   Sheikh, K., et al., "Smart Antennas for Broadband Wireless Access Network," *IEEE Communications Magazine*, Vol. 37, No., 11, pp. 100–105.

[74]   J. Bach Andersen, "Role of Antennas and Propagation for the Wireless System Beyond 2000," *Wireless Personal Communications*, Vol. 17, No. 2–3, pp. 303–310.

[75]   Tarok, V., N. Seshandri, and R. C. Calderbank, "Space-Time Codes for High Data Rate Wireless Communication: Performance Criteria and Code Construction," *IEEE Transaction on Information Theory*, Vol. 44, No. 2, 1998, pp. 744–765.

[76]   Nanda, S., K. Balachandran, and S. Kumar, "Adaptation Techniques in Wireless Packet Data Services," *IEEE Communications Magazine*, Vol. 38, No. 1, 2000, pp. 54–64.

[77]   Berrou, C., A. Glavieux, and P. Thitimajshima, "Near Shannon Limit Error-Correcting Coding and Decoding: Turbo-Codes," Proc. *ICC93*, May 1993.

[78]   Benedetto S., and G. Montorsi, "Unveiling Turbo Codes: Some Results on Parallel Concatenated Coding Schemes," *IEEE Transactions on Information Theory*, Vol. 42, No. 2, March 1996, pp. 409–428.

[79]   Robertson, P., and T. Worz, "A Novel Bandwith Efficent Coding Scheme Employing Turbo Codes," *Proc. ICC96*, June 1996.

[80]   Benedetto, S., et al., "Parallel Concatenated Trellis Coded Modulation," *Proc. ICC96*, June 1996.

[81]   Benedetto, S., et al., "Serial Concatenated Trellis Coded Modulation with Iterative Decoding: Design and Performance," *Proc. Comm. Theory Miniconf. 97*, Nov. 1997.

[82]   Mitchell, T., "Broad Is the Way [Ultra-Wideband Technology]," *IEE Review*, Vol. 47, No. 1, Jan 2001, pp. 35–39.

[83]   Verdù, S., *Multiuser Detection*, Cambridge, UK: Cambridge University Press, 1998.

[84] Buzzi, S., M. Lops, and A. M. Tulino, "MMSE Multi-User Detection for Asynchronous Aual Rate Direct Sequence CDMA Communications," *Proc. 9th PIMRC*, Boston, Massachusetts, Sept. 1998.

[85] Mitra, U., "Comparison of Maximum Likelihood-Based Detection for Two Multi-Rate Access Schemes for CDMA Signals," *IEEE Transactions on Communications*, Vol. 47, Jan. 1999, pp. 64–77.

[86] Buzzi, S., M. Lops, and A. M. Tulino, "Blind Adaptive MMSE Detection for Asynchronous Dual-Rate CDMA Systems: Time-Varying Versus Time-Invariant Receivers," *Proc. IEEE GLOBECOM'99*, Dec. 1999.

[87] Bensley, S.E., and B. Aazhang, "Subspace-Based Channel Estimation for Code-Division Multiple-Access Communication Systems," *IEEE Transactions on Communications*, Vol. 44, Aug. 1996, pp. 1009–1020.

[88] Weiss, A. J., and B. Friedlander, "Channel Estimation for DS/CDMA Downlink with Aperiodic Spreading Codes," *IEEE Transactions on Communications*, Vol. 47, Oct. 1999, pp. 1561–1570.

[89] Ramakrishna S., and J. M. Holtzman, "A Scheme for Throughput Maximization in a Dual-Class CDMA System," *IEEE Journal on Selected Areas on Communications*, Vol. 16, Aug. 1998, pp. 830–844.

[90] Dziong Z., M. Jia, and P. Mermelstein, "Adaptive Traffic Admission for Integrated Services in CDMA Wireless-Access Networks," *IEEE Journal on Selected Areas on Communications*, Vol. 14, Dec. 1996, pp. 1737–1747.

[91] Soroushnejad, M., and E. Geraniotis, "Multi-Access Strategies for an Integrated Voice/Data CDMA Packet Radio Network," *IEEE Transactions on Communications*, Vol. 43, Feb./Mar./Apr. 1995, pp. 934–945.

[92] Caceres, R., and V. N. Padmanabhan, "Fast and Scalable Handoffs for Wireless Internetworks," *Proc. of ACM MobiCom'96*, Nov. 1996.

[93] Benvenuto, N., and F. Santucci, "A Least Squares Path Loss Estimation Approach to Handover Algorithms," *IEEE Transactions on Vehicular Technology*, Vol. 48, March 1999.

[94] Graziosi, F., and F. Santucci, "Analysis of a Handover Algorithm for Packet Mobile Communications," *Proc. ICUPC'98*, Florence, Italy, Oct. 1998, pp. 769–774.

[95] Efthymiou, N., Y. F. Hu, and R. E. Sheriff, "Performance of Intersegment Handover Protocols in a Integrated Space/Terrestrail-UMTS Environment," *IEEE Transactions on Vehicular Technology*, Vol. 47, No. 4, Nov. 1998, pp. 1179–1199.

[96] Maral, G., et al., "Performance Analysis of a Guaranteed Handover Service in a LEO Constellation with a Satellite—Fixed Cell System," *IEEE Transactions on Vehicular Technology*, Vol. 47, No. 4, Nov. 1998, pp. 1200–1213.

[97] Lilleberg, J., and R. Prasad, "Research Challenging for 3G and Paving the Way for Emerging New Generation," *Wireless Personal Communications*, Vol. 17, No. 2, June 2001, pp. 355–362.

2

Multiple Access Protocols

2.1 Introduction

A multiple access protocol (MAP) establishes a method of sharing common resources. In case of communication systems the resource to be shared is the communication channel. In absence of such a protocol, conflicts occur if more than one user tries to access the channel at the same time. A MAP should avoid or at least resolve these conflicts. Of course, one could avoid the need for a MAP by letting each user have its own resources. However, there are two good reasons to share resources among users. First, resources may be scarce and/or expensive. This is the case of a mainframe computer usually time-shared among a number of users, but it is also the case of wireless communication systems where the common and unique resource to be shared is the ether. Second, there may be the need for a user to be able to communicate with all the other users. An example of this case is the telephone system, wireless or fixed, where users share common switching centers, thus a connection can be made between any two users.

MAPs are a subset of the MAC protocols. These protocols belong to layer 2 of the International Standard Organization/Open Systems Interconnect (ISO/OSI) reference model, the data link control layer, whose main tasks include accessing the medium, multiplexing of different data streams, correction of transmission errors, synchronization, and implementation of reliable point-to-point/point-to-multipoint connections.

The aim of this chapter is to provide the reader with an updated vision on MAPs, focusing attention on the wireless domain and multimedia traffic. All the MAPs adopted in current wireless communication systems are an adapted version of traditional and well-known mechanisms designed for wired networks or for specific wireless scenarios and applications, such as satellite maritime communications and circuit-switched (CS) voice. While fixed-assignment protocols may be suitable for CS voice services, they are not suitable for bursty data traffic or for systems that have to support both voice and data communications. Random protocols or demand-assignment-based schemes are more efficient in these cases. Furthermore, in order to support traffic with variable bit rates, different QoS requirements, other concepts like priority, scheduling, and QoS support should be taken into account for the design of MAPs. This evolution of MAPs is summarized in Figure 2.1.

Basic multiple access mechanisms are introduced in Section 2.3, according to the classification of MAPs provided in Section 2.2 in random access protocols (e.g., ALOHA), contentionless protocols (frequency/time division multiple access, PRMA) and CDMA-based protocols. Note that these multiple access protocols have been extensively dealt with in other books [1–9]. Sections 2.3 through 2.5 provide basic concepts for a better understanding of the rest of the chapter. Section 2.6 describes the evolution of these basic mechanisms to more complex hybrid implementations, gradually adapted to current wireless communication systems. A description of protocols that are designed for specific wireless scenarios (micro- macrocellular scenarios or specific physical layer interfaces) and that guarantee a more flexible management of different types of traffic and

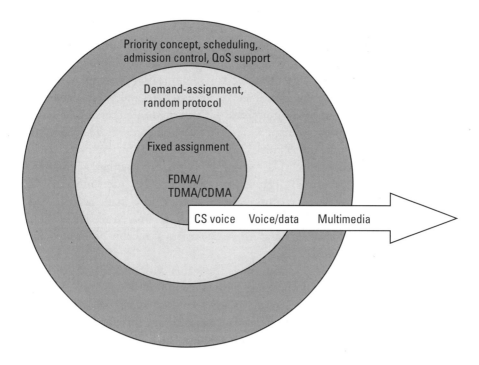

Figure 2.1 Evolution of MAPs.

QoS requirements, is provided. In particular, MAC protocols that are being proposed for supporting real and non-real-time services in the standard WCDMA and UTRA TDD are presented. Section 2.6 also highlights design guidelines of MAPs suitable for wireless multimedia traffic.

2.1.1 Desired Properties

2.1.1.1 General properties

The design of a protocol is usually accomplished with a specific environment in mind: the properties of the protocol are then mainly determined by the design goals. If we rule out the environment-specific properties, a number of properties that any good MA protocol should possess can be still identified. These properties are as follows:

1. The protocol should perform the allocation of the channel capacity among the users to use the transmission medium efficiently. The efficiency is usually measured in terms of channel throughput and transmission delay.

2. The allocation should be fair towards each user. This means that each user should (on the average) receive the same allocated capacity, unless priorities are assigned to the users.

3. The protocol should be flexible in allowing different types of traffic (e.g., voice and data).

4. The protocol should be stable. This means that if the system is in equilibrium, an increase in load should move the system to a new equilibrium point. With an unstable protocol, an increase in load will force the system to continue to drift to a higher load and a lower throughput.

5. The protocol should be robust with respect to equipment failure and changing conditions. If one user does not operate correctly, this should affect the performance of the rest of the system as little as possible.

2.1.1.2 Wireless Scenario

In the wireless scenario, the protocol should be able to deal with the following:

1. *The hidden terminal problem:* Two terminals are out-of-range (hidden from) of each other by a hill or by a building or by some physical obstacle opaque to signals of interest but both are within the range of the central or base station.

2. *The near-far effect:* Transmissions from distant users are more attenuated than transmissions from the closest users.

3. *The effects of multipath fading and shadowing experienced in radio channels.* In particular, a multipath medium with a large delay spread relatively to the transmission bit rate is characterized by inter-symbol interference (ISI). The output of the receiving filter, sampled at the times $t = kT + \tau_0$, where T is the symbol interval and $\tau_0 < T$, has the following structure:

$$y_k = g_k + \sum_{\substack{n=0 \\ n \neq k}}^{\infty} g_n x_{k-n} + v_k \qquad (2.1)$$

where g_k represents the desired information symbol at the kth sampling instant, x_{k-n} and v_k are the sampled pulse representing the response of the receiving filtering to the input impulse and to the additive Gaussian noise, respectively. The second term in (2.1) represents the ISI. The effect of ISI is to increase the possible errors caused by additive noise. It also distorts the position of the zero-crossings and causes the system to be more sensitive to a synchronization error.

4. The effects of cochannel interference in cellular wireless systems caused by use of the same frequency band in different cells.

Most of the above-mentioned protocol properties are in conflict, and a trade-off is needed in the protocol design. Furthermore, if the scenario of interest is multimedia traffic on a wireless channel, other requirements should be taken into account in the design of multiple access protocols. The specific design of MAPs for this scenario is discussed in Section 2.6.

2.2 Classification of MAPs

MAPs can be classified into three groups [2, 3]: *contention* (or random) protocol, *contentionless* (or scheduling) protocols, and *CDMA* protocols.

With contention protocols, a user cannot be sure that a transmission will not collide because other users may be transmitting at the same time. They can be further subdivided into *repeated random* protocols and random access protocols with reservation. With the former protocols, with every transmission there is the possibility of contention. With the latter protocols, only in its first transmission does a user not know how to avoid collisions with other users. Once the user has accessed the channel, however, further transmissions of that user are scheduled until completion of the user transmission. Explicit reservation type protocols use a short reservation packet to request transmission at scheduled times. Implicit reservation protocols are designed without the use of any reservation packet.

Contentionless protocols avoid that the possibility of two or more users accessing the channel at the same time by scheduling the transmissions of the users. This can either be done in a fixed-assignment fashion, where each user is allocated part of the transmission capacity, or in a demand-assignment fashion where scheduling only takes place between the users that have something to transmit.

CDMA protocol falls between the other two categories. In principle, it is a contentionless protocol where a number of users are allowed to transmit simultaneously without conflict. However, if the number of simultaneously transmitting users rises above a threshold, contention will occur.

According to the above classification, random protocols are introduced in Section 2.3. Then, Section 2.4 introduces contentionless protocols, starting with fixed-assignment protocols like TDMA and FDMA, followed by the demand-assignment based schemes. In Section 2.5, CDMA-based protocols are described. In particular, insights on direct-sequence CDMA schemes and some hybrid TDMA/CDMA schemes are provided.

2.3 Random Access Protocols

With random MAC protocols, there is no central or distributed scheduling of transmissions. A user who is ready to transmit is not aware of the optimal starting time to avoid interference with other transmitting users. Even if the user could be aware of ongoing transmissions by sensing the channel, he has no information about other users ready to start an almost simultaneous transmission. If several users start an almost simultaneous transmission, a *collision* occurs, and all the overlapping packets are lost. The contention can be solved by the protocol or by higher layers (e.g., retransmission of data). The simplest version of this class of protocols is the so-called pure-ALOHA (p-ALOHA). It was conceived at the University of Hawaii and it was used in 1970 in the ALOHANET for wireless connection of several stations [10]. Several changes to this basic scheme have been proposed to increase the efficiency and/or to render it more suitable to scenarios other than the original one. Most of the existing MAC protocols include an ALOHA component.

2.3.1 p-ALOHA

Each user can access the medium at any time. ALOHA neither coordinates medium access nor resolves contention on the MAC layer. If a collision occurs, the user will retransmit the packet after a random amount of time. Waiting a random time before retransmitting mostly avoids the chance of users who collide the first time will collide again during retransmissions.

In case of a centralized protocol, upon reception of correct transmission, the base station sends an acknowledgment packet on the downlink channel. When receiving the acknowledgment, the user knows transmission has been successful. The base station recognized the occurrence of a collision since it receives a garbled transmission. The time diagram of the described protocol is shown in Figure 2.2.

In case of a distributed protocol, the user "listens" to the channel and checks if there are any other ongoing transmissions.

This very simple scheme is effective for a light load. The probability that a user transmission is interrupted by other transmissions becomes quite large if the traffic on the channel increases. As is shown in Figure 2.2, if user Y starts transmission at $t = t_0$ and the transmission lasts T_s seconds, the transmission of a user starting anytime within the time period between $(t_0 - T_s)$ to $(t_0 + T_s)$ collides with the transmission of user Y. The vulnerable period is $2T_s$ and the probability that the transmission of the packet is successful is the probability, denoted by p, that no packet transmission begins during the time interval $2T_s$. Let us denote with G the total rate of transmission attempts, in packets per time slot. Assuming a Poisson distribution for the arrival of packets, this probability is equal to

$$p = e^{-2G} \tag{2.2}$$

The rate S of successful transmissions, also in packets per time slot, is given by

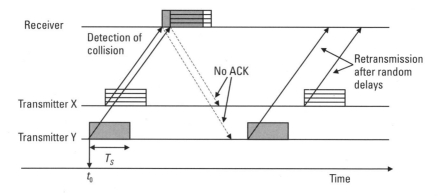

Figure 2.2 Diagram illustrating the principle of the p-ALOHA random access protocol with collision and centralized control.

$$S = Gp = Ge^{-2G} \qquad (2.3)$$

The maximum value of the right-hand side of (2.3) is $1/2e = 18\%$. If nodes randomize their transmissions suitably, then S approaches the upper bound that represents the efficiency of the p-ALOHA protocol [11].

Energy-consumption is also an important issue in the design of MAC. Portable wireless devices have severe constraints on the energy consumption and, hence, on the average transmission power level instead of peak power level. The same can be said for the downlink of a satellite communication system. In these cases, the channel average data rate for a fixed average transmission power and a fixed bandwidth is a more appropriate figure of merit. In [11], the following definition of efficiency is proposed for the ALOHA protocol:

$$\varepsilon = \frac{Ge^{-2G} \log\left(1 + \dfrac{P}{GN}\right)}{\log\left(1 + \dfrac{P}{N}\right)} \qquad (2.4)$$

where the ratio P/G is the average power during the transmission of a packet when the ALOHA protocol is considered; P/N is the signal to noise power ratio at the receiver (Gaussian noise is assumed) for a continuous channel using the same average transmission power and the same total bandwidth. The efficiency of p-ALOHA approaches 1 when small values of throughput and of signal-to-noise ratio are considered.

2.3.2 Slotted-ALOHA

A first refinement to the p-ALOHA scheme is provided by the introduction of time slots. All transmissions can start at the beginning of a time slot. As a result, the vulnerable period of a transmission is shorter than in a p-ALOHA. If user Y generates a packet between time 0 and T_s, the transmission of this packet is delayed until the time $t = T_s$. Only users generating packets between time 0 and T_s will transmit at the same time $t = T_s$ and collide with user Y. As shown in Figure 2.3, the vulnerable period of a transmission is only T_s and it is halved compared to p-ALOHA. This doubles the efficiency to 36%.

2.3.3 Carrier Sense Multiple Access

A further improvement to the basic ALOHA is sensing the carrier to know if there are ongoing transmissions [12, 13]. The transmission is started only if the carrier is idle (i.e., no other transmissions are ongoing). Due to the propagation

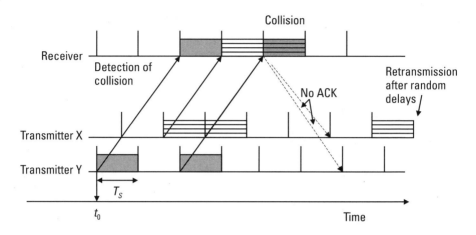

Figure 2.3 Diagram illustrating the principle of the slotted-ALOHA random access protocol with collision and centralized control.

delay between two users, it is possible for a user to sense the idle channel and start its transmission while another transmission is already in progress, thus causing a collision. A user is informed of a collision by the absence of an acknowledgment packet from the receiving station. Upon detecting a collision, the packet is rescheduled for transmission a random time later. This class of protocols is named carrier sense multiple access (CSMA). The main subclasses are as follows: nonpersistent CSMA, p-persistent CSMA, CSMA with collision detection (CSMA/CD).

In the nonpersistent CSMA, a user that has to transmit a packet "listens" to the channel for ongoing transmissions, starting the transmission immediately if the channel is idle; if the channel is busy, the user waits a random time before sensing the channel again.

In p-persistent CSMA systems, when the channel is busy, the user keeps sensing the channel until it becomes idle. Then it transmits with probability p and defers the transmission with probability $(1 - p)$ of t seconds, with t being the maximum propagation delay between any two users in the system. After t seconds the deferred terminal senses the channel again and applies the same algorithm. In case of 1-persistent CSMA, all the users that become ready to transmit when the channel is busy transmit as soon as the channel becomes idle thus colliding each other. General p-persistent algorithms randomize the transmission times of the accumulated packets and hence reduce the probability of collisions.

With CSMA/CD, a user keeps monitoring the channel while transmitting. In case of collision, the user stops the packet transmission, waiting for a random time before trying again. This results in a reduction of the time a user needs to react in case of collision, and hence, in an increased efficiency of the algorithm. Without collision detection mechanisms, the user learns about a collision after its whole packet has been sent and received. Even with collision detection, stations may be transmitting for a while before noticing packet collisions. The wasted time increases with the propagation time. This scheme is used by the LANs following the IEEE 802.3 standard. The efficiency, as defined in Section 3.2, can be written as [6]

$$\eta_{CSMA/CD} \approx \frac{1}{1+5a} \quad \text{with} \quad a := \frac{\rho}{\tau} \qquad (2.5)$$

where τ is the packet transmission time and ρ is the maximum propagation time across the shared transmission channel. For a 10BASE-T network, one of the most commonly used versions of IEEE 802.3 networks, the efficiency ranges typically between 40% and 90%. It increases with the average packet length and decreases with the packet propagation time and the transmission rate.

CSMA protocols suffer from the problem of *hidden and exposed terminals*, which is peculiar of wireless networks [13]. In Figure 2.4, four wireless nodes are considered and their transmission range is represented by the dashed ellipses.

In the figure, B is the sender, C is the receiver, A (the so-called *exposed terminal*) is in the range of the sender but not of the receiver, and D (the so-called *hidden terminal*) is in the range of the receiver but not of the sender. While B is transmitting to C, D might start transmitting since it is unable to "hear" the transmission even if it senses the channel. Thus, a collision may occur. Furthermore, since B is unable to detect this collision, it keeps on transmitting. On the other hand, while B is transmitting to C, if A aims at transmitting to another node outside the areas covered by B and C, it senses the channel and detects that the carrier is busy, hence deferring transmission. In this case, since C is outside the range of A, deferring transmission is not necessary. In both cases, the protocol is not efficient. A proper power control that changes the transmission ranges of each terminal could help in reducing the impact on the throughput of these effects [14].

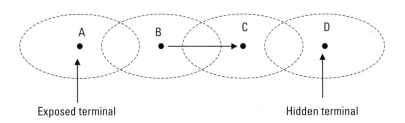

Figure 2.4 Hidden and exposed terminals with CSMA protocols.

In spite of the hidden and exposed terminal problem, several forms of this basic scheme are used in most WLANs (e.g., HIPERLAN, IEEE 802.11). The IEEE 802.11 standard uses CSMA with collision avoidance (CSMA/CA). The latter protocol is similar to CSMA/CD, but it allows smoother competition among the stations by favoring stations that happened to wait longer. The efficiency of the scheme depends on propagation delay and packet size.

2.3.4   Inhibit Sense Multiple Access

The class of Inhibit Sense Multiple Access (ISMA) protocols or Busy Tone Multiple Access (BTMA) represents a solution to the problem of hidden terminals [15–18]. They differ from CSMA protocols only in the way of sensing the channel. In CSMA sensing takes place by listening to the channel where users transmit. In ISMA a base station transmits a busy/idle signal on a separate outbound channel to highlight presence or absence of a transmission in progress. As soon as a base station receives a signal from a user on the inbound channel, it will generate a busy signal on the outbound channel. When the transmission ends, the base station will transmit an idle signal. If two users are hidden from each other but not from the base station they will still be able to determine if the other user is transmitting or not. As in the case of CSMA protocols, ISMA protocols can be classified as nonpersistent, p-persistent, and ISMA with collision avoidance (ISMA/CD). This class of protocols is used in the IS-136 system for the packet data transmission service cellular digital packet access (CDPD), also known as Digital Sense Multiple Access (DSMA).

2.3.5   Capture Effect

In a wireline environment, collisions occur whenever the number n of packets that overlap in time at the receiver is such that $n \geq 2$. In a wireless environment, fading can suppress some of the user signal levels, and the radio receiver is able to be captured by the strongest of the overlapping packets and thus receive this packet correctly. Therefore, some of the colliding packets might not be lost and efficiency evaluations of MAPs that do not take this effect into account are pessimistic.

Let us assume that there is a test packet j and n interfering packets. The capture probability is defined as

$$P_{cap} = \text{prob}\left(P_{R,j} \geq z\left\{ \sum_{i=1, i \neq j}^{n} P_{R,j} + N \right\} \right) \qquad (2.6)$$

where $P_{R,i}$ is the received power at the base station due to the transmitter i; N is a nonnegative random variable that represents the effect of additive noise, such as receiver noise or interference from transmitter in other systems; z is the power ratio threshold of the capture model, which is the minimum carrier-to-interference ratio (CIR) needed for successful reception, and it is determined, for instance, by the modulation type and the receiver sensitivity. For typical narrowband systems, $1 < z < 10$ dB. For a direct-sequence spread-spectrum system, the processing gain effectively reduces the effect of interference from other users, so the value of z is roughly inversely proportional to the processing gain. For such systems, typically $0.1 < z < 1$ dB [19].

Techniques to purposely induce or enhance capture effect are discussed in [20].

The following section provides a throughput analysis of *unslotted np-ISMA* protocols with capture effect.

2.3.5.1 Throughput Analysis of Unslotted np-ISMA with Capture

The probability U that the channel, during a cycle (i.e., an idle period plus a transmission period), is used without conflicts is defined as the probability that a test packet overlapped by n interfering packets, multiplied by the probability that the test packet is received correctly in the presence of n interfering packets; that is,

$$U = \sum_{n=0}^{\infty} P_n(n)(n+1)P_{cap} \tag{2.7}$$

with

$$P_n(n) = \frac{(dG)^n}{n!} e^{-(dG)} \tag{2.8}$$

where d is the transmission delay of the inhibit signal necessary to switch from "busy" to "idle."

The term $(n + 1)$ is included in (2.7) because when inhibit bits are in the outbound channel and the test packet is subjected to n interfering packets, $(n + 1)$ packets are presented during the busy period. One of these $(n + 1)$ packets is capable of capturing the receiver with a certain probability. The throughput for unslotted nonpersistent ISMA with capture can then be found by dividing U by the sum of the expected duration of the transmission period T, given by

$$T = 1 + 2d - \frac{1}{G}\left[1 - e^{(-dG)}\right] \tag{2.9}$$

and the expected duration of the idle period I, with $I = 1/G$.

Thus, the throughput of the unslotted nonpersistent ISMA is

$$S = \frac{U}{T+I} = \frac{\sum_{n=0}^{\infty} \frac{(dG)^n}{n!} e^{(-dG)}(n+1)}{(1+2d) + \frac{1}{G} e^{(-dG)}} P_{cap} \qquad (2.10)$$

Studies on P_{cap} in presence of shadowing and different kinds of fading can be found in [3, 21–25]. Figures 2.5 and 2.6 compare the throughput performance of slotted ALOHA and unslotted nonpersistent ISMA protocols in a Rician fading channel with receiver capture.

It can be seen from these figures that the channel capacity of slotted-ALOHA protocols in a Rician fading channel is higher than 0.36 (the channel capacity of ideal slotted-ALOHA). Furthermore, it can be seen that the channel capacity of ISMA with inhibit delay factor $d = 0.5$ is lower than the channel capacity of ALOHA while the capacity of ISMA with $d = 0.05$ is higher than that of ALOHA. It is worth mentioning here that the inhibit delay factor is low if the cell size is low. Therefore, the unslotted ISMA protocol would be a better choice for indoor wireless communications than the slotted-ALOHA protocol.

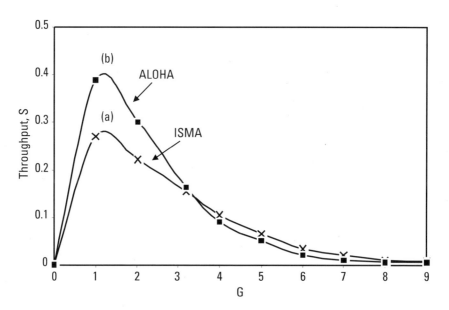

Figure 2.5 Throughput curves for an inhibit delay factor of 0.5 for (a) unslotted ISMA with a Rician parameter of 7 dB and z = 3 dB; and (b) slotted ALOHA with a Rician parameter of 7 dB and z = 3 dB.

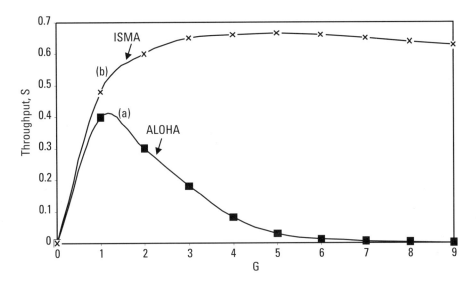

Figure 2.6 Throughput curves for an inhibit delay factor of 0.05 for (a) slotted ALOHA with a Rician parameter of 7 dB and $z = 3$ dB; and (b) unslotted ISMA with a Rician parameter of 7 dB and $z = 3$ dB.

2.4 Contentionless MAPs

The contentionless multiple access protocols avoid multiple users trying to access the same channel at the same time by scheduling the transmissions of all users. The users transmit in an orderly scheduled manner, and hence, every transmission is successful. Either a *fixed-assignment* scheduling or a *demand-assignment* can be adopted.

2.4.1 Fixed-Based Assignment Protocols

Fixed-assignment protocols assign each user a predetermined and fixed allocation of resources, regardless of the user's need to transmit. This might be appropriate for voice traffic, but could be quite wasteful for bursty data traffic. An allocation in the frequency domain results in an FDMA protocol. An allocation in the time domain results in a TDMA protocol. Hybrid FDMA/TDMA are commonly adopted. The basic concepts on these multiple access protocols will be recalled in the next two subsections.

2.4.1.1 FDMA

In FDMA, the system bandwidth is divided into orthogonal channels with nonoverlapping frequencies. It can be seen from Figure 2.7(a) that each user is allocated a unique frequency band or channel.

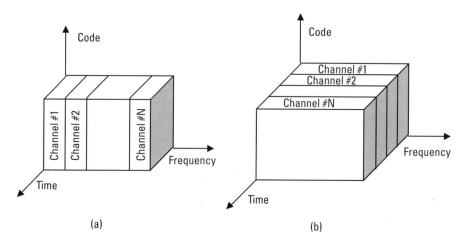

Figure 2.7 Concept schemes of (a) FDMA; and (b) TDMA.

B being the bandwidth of each subchannel, ideally N users can be simultaneously supported with a total bandwidth of BN. Due to the nonideal passband filters at the receiver, guard frequency bands are needed between two consecutive channels to reduce the adjacent-channel interference. That approach increases the required bandwidth and reduces the spectral efficiency of the network. Therefore, FDMA is usually implemented either in narrowband systems (where B is relatively small; e.g., 30 kHz) or to produce few subchannels (small values of N), combined with other multiple access techniques (e.g., TDMA, CDMA). In narrowband FDMA systems, the symbol time is usually large as compared with the average delay spread, and hence, the amount of ISI is low and no equalization is required. Furthermore, few bits are needed for overhead purposes such as synchronization and framing as compared to TDMA.

FDMA systems have to cope with intermodulation (IM) products interference. The latter are due to the nonlinearity of power amplifiers or power combiners when operated at or near saturation for maximum power efficiency, which is the typical situation in mobile radio systems. Passing a signal that is the sum of N frequencies through a power amplifier or a hard limiter will produce (usually) undesired harmonics outside and inside the mobile radio band. Harmonics out of band can cause interference to services that are using adjacent frequencies. Inband harmonics cause interference to other users in the mobile systems using adjacent frequencies. These harmonics are IM frequencies. The IM product spreading can be reduced to a desired low value by a proper channel frequency assignment. Anyway, there is a trade-off between the amount of IM noise the system can tolerate and the required bandwidth.

Real systems almost always include an FDMA component. In cellular systems, the two directions, base to mobile station and vice versa, are usually separated in frequency. This scheme is called FDD. Of course, both receiver and transmitter have to know the frequencies in advance since the receiver must be able to tune properly. It is not possible to jump arbitrarily in the frequency domain, contrary to what can be done in the time domain.

2.4.1.2 TDMA

Time is divided into nonoverlapping time slots [Figure 2.7(b)]. The transmission from various users is organized in a repeating frame structure. N time slots comprise a frame and each user occupies cyclically a time slot of this frame. A physical channel in a TDMA system is defined as a time slot that reoccurs each frame. Therefore, as the transmission of a stream is noncontinuous in time, digital data modulation must be used unlike in FDMA systems that can accommodate analog frequency modulation (FM). Base stations solve the competition between different mobile stations that want to access the medium by allocating time slots for channels in a fixed pattern. The crucial factor concerns accessing the reserved time slot at the right moment. TDMA schemes with fixed access patterns guarantee a fixed transmission delay and are used for many second generation digital mobile phone systems like IS-54, IS-136, and GSM.

The GSM system deploys TDMA combined with slow frequency hopping (FH). The principle of FH is that each TDMA burst is transmitted via a different RF channel: as a consequence, if one burst happens to be in a deep fade, the next burst most probably will not be. Therefore, the physical channel is defined as a sequence of radio frequency channels and time slots. Each carrier frequency supports eight physical channels mapped onto eight time slots within a TDMA frame, as shown in Figure 2.8.

Each time slot consists of 156.25 bits, out of which 8.25 bits are used for guard time intervals. The latter are provided to prevent burst overlapping due to propagation delay fluctuations, and six are start and stop bits that are used to prevent overlap with adjacent time slots. Each time slot is 0.577 ms. The 13th and 26th frames in the multiframe structure are used only for control purposes.

TDMA has the advantage that it is possible to allocate different numbers of time slots per frame to different users. Thus, bandwidth can be supplied on demand either to different users or to the uplink and downlink of the same user, in case of TDD mode, by concatenating or reassigning time slots based on priority.

Interference and Power Management

With TDMA, each user utilizes the whole available bandwidth, and hence, there is an increased probability that multipath induces ISI at high bit rate. Due to the time-varying nature of the multipath-fading channel, users experience a different channel impulse response at every TDMA burst. The insertion of a known training sequence in each slot allows the receiver to learn the channel impulse

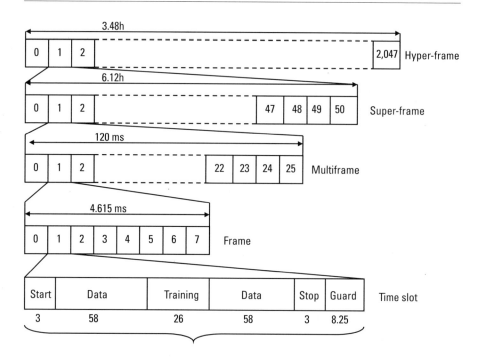

Figure 2.8 Frame and time slot structure in GSM.

response and to adapt its filter coefficients accordingly. Furthermore, adaptive equalizers are needed since the channel can change appreciably during a single TDMA burst. Single-carrier TDMA solutions are limited in supportable transmission rates by equalizer complexity. Even though new techniques such as interference suppression and space-time processing are promising, the interactions of these techniques with equalization significantly lower achievable bit rates in hostile operating environments for single-carrier solutions.

In TDMA cellular systems, available channels are reused in different sets of cells called *clusters*. Cells belonging to different clusters can use the same channels, thus causing cochannel interference. The spatial separation of cells that reuse the same channel set is called *reuse distance* and should be large enough to reduce the interference and as small as possible to maximize the spectral efficiency. Therefore, frequency planning is a fundamental issue to increase the spectral efficiency and the capacity.

There are three basic types of channel assignments: *fixed* channel assignment (FCA), *dynamic* channel assignment (DCA), and *flexible* channel assignment. With FCA, each cell has a fixed subset of the channels that can be assigned [8]. Frequency planning with FCA was used in first generation analog FDMA mobile systems. With DCA, all the channels available can be dynamically assigned to a cell, according to the traffic load. Some second generation systems have already deployed simple forms of DCA, but generally as a means of automatic channel assignment or capacity enhancement, but just at local level. The reason for not fully exploiting the potential capacity gain of DCA are the difficulties introduced by excessive exchange of channel-usage information for coordination among cells, and intensive receiver measurement required by high-performance DCA algorithms [26]. As a result, the DCA gain is limited to somewhat better traffic resource utilization. FCA and DCA can be combined, assigning each cell with a fixed set of channels, while a pool of channels is reserved for flexible assignment (e.g., channel borrowing [27–29]).

In FDMA/TDMA systems, cell planning is based on the worst-case situation where users at the cell boundary can still have acceptable signal quality. Such an arrangement is simple but inefficient, because a user close to the base station transmits at unnecessarily high power levels, which not only results in high cochannel interference, but also reduce the life-time of the battery in the mobile unit. Transmitter power control is exploited to solve this problem [30], by dynamically adjusting transmitter power levels to minimize the interference levels at the receiver, still providing each user with adequate quality. Transmitter power control is an integral part of almost all 2G cellular systems. Efficient interference management aims at achieving acceptable CIR ratios in all active communication links in the system. For GSM, the mobile has a transmit power dynamic range of about 30 dB (i.e., 20 mW to 20W), which can be adjusted through adaptive power control at a rate of 2 dB every 60 ms to achieve constant received signal power at the base station.

Other techniques that can be applied to reduce the cochannel interference in TDMA cellular systems are proper handover procedures and slow frequency hopping. A handover strategy that selects the BS with equal or higher received signal strength even if the received signal level from the current BS is sufficient to support the call, is better from both the interference management and power consumption point of view. For GSM, the handover strategy is based on this concept.

Slow frequency hopping among carrier channels is a useful diversity method against Rayleigh fading, as it reduces the probability that the received signal will fall into a deep fade for a long duration. MS and BTS may change the carrier frequency after each frame based on a common hopping sequence. An MS changes its frequency between uplink and downlink slots. This is especially useful for a slowly moving or stationary mobile which otherwise would experience deep fades for a relatively long time.

2.4.2 Demand-Based Assignment Protocols

Demand-based assignment protocols assign resources according to reservations submitted by the users. Requests are transmitted using dedicated or random access channels and are then processed. According to the processing results, users are or are not assigned part of the channel capacity. This process can be controlled by a central entity as in *centralized* demand-based assignments protocols, or can be distributed among the users as in *distributed* demand-based assignments protocols. Those schemes allow higher throughput than fixed-assignments schemes in case of hybrid real-time/non-real-time multimedia traffic. The main problems of those protocols are the additional overhead and delay caused by the reservation process also under light load.

2.4.2.1 Demand Assigned Multiple Access

The Demand Assigned Multiple Access (DAMA) belongs to the category of reservation ALOHA (r-ALOHA) schemes. The INMARSAT network, employing a simple ALOHA random access request channel for a large number of potential ship stations, was the first commercial demand-assignment network with random access request [31, 32]. According to the protocol, the system, as shown in Figure 2.9, has two operation modes: reserved mode and ALOHA mode.

If none of the time slots is allocated because there are no reserved users, the system is in the ALOHA mode. During that phase all the stations can try to reserve future slots by sending requests in one of the V mini-slots that compose the frame. The contention of the mini-slots follows a slotted-ALOHA mechanism. As soon as a user successfully transmits a reservation request in one of the mini slots, the system switches to the reserved mode. In the reserved mode the frame is divided in M + 1 time slots, where M slots or less are assigned to reserved users. The last slot is divided in V mini-slots used for reservation requests. A broadcast acknowledgment packet informs the user about its

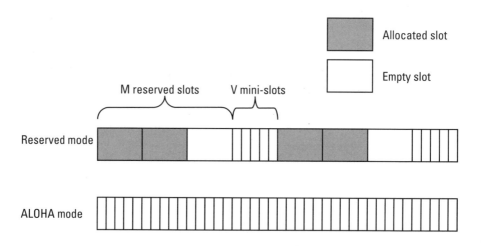

Figure 2.9 Reserved state and ALOHA state in DAMA protocol.

allocated full time slot. Each reserved user is allocated one or more of the M slots. DAMA is an *explicit reservation* scheme since each transmission slot has to be reserved explicitly.

2.4.2.2 Roll-Call Polling

The Roll-Call Polling is a centralized demand assignment protocol. A central controller polls the users one by one in a predetermined sequence. Each user is polled only once in a sequence. Upon reception of a polling message from the central controller, a user either transmits a control message indicating that there is no need to transmit, or it transmits all messages that have accumulated in its buffer. After transmitting the last packet, the user sends a "ready" message to the controller. Upon reception of this message, the central controller sends a polling message to the next user in the sequence.

The overhead in the Roll-Call Polling consists of the polling messages and the control messages stating that a user has nothing to transmit or has finished the transmission. The overhead grows with the number of users that have nothing to transmit. More advanced polling protocols have been developed to reduce the overhead [33]. Some polling protocols also allow users generating traffic to be polled more than once during a sequence. Similar schemes are used in Bluetooth and represent one possible access scheme in IEEE 802.11.

2.4.2.3 Token-Passing Protocol

The Token-Passing Protocol is widely used for LANs. This protocol belongs to the category of distributed demand assignment protocols, where all users are involved in the scheduling process. Each user is given a unique address and the users of the bus are ordered in a cyclic fashion, hence forming a logical ring. A specific bit pattern, called the token, circulates on the ring. A user that receives the token is allowed to transmit. After the transmission, the user sends the token to the successor in the ring. If the user has nothing to transmit, it will transmit the token to its successor either immediately or after a time interval. For instance, in the IEEE 802.5 every station can hold the token for up to 10 ms and transmit a number of packets back to back before releasing the token. Further details on MAC protocols in the IEEE 802.5 standard can be found in [6].

2.4.2.4 Packet Reservation Multiple Access

The Packet Reservation Multiple Access (PRMA), originally proposed by Goodman et al. [34], is a demand-based assignment scheme with slotted-ALOHA for reservation contention periods and *implicit reservation*. It represents an adaptation to the cellular environment of the r-ALOHA, conceived for satellite communications [32]. A certain number of slots forms a frame that is repeated in time. A base station broadcasts the status of each slot (idle/busy) to all mobile stations. Packets generated by the user are stored in a first-in-first-out buffer, and the user competes for free slots in ALOHA fashion to transmit the first packet from the buffer. Upon a successful transmission of the first packet, all future slots are *implicitly* reserved for this station. As soon as the user interrupts its transmission, it loses its reservation of the time slot. In this way, slots left unused, for instance due to silence gaps of a speech user or due to the fact that a packet data user temporarily has no data to transmit, can be used by other users. Therefore, PRMA exploits the condition that at any given instant a substantial number of users are not talking. Since there are more users than channels, there will be times in which not enough channels are available and the quality of some channels becomes severely degraded. To maintain low subjective speech degradation, the packet delay due to the queuing must be lower than 32 ms. Therefore, in PRMA if the first packet has not been transmitted within 32 ms, the packet is dropped from the buffer and the user tries to transmit the next packet. The dropping probability due to the contention for free slots must be less than 1%. Among active terminals, which have obtained access to the system through a successful transmission of the first packet, the system operates like a fixed TDMA and ensures transmission with guaranteed data rate. This mechanism is suitable to the voice, which, in a talkspurt, is generated at a constant rate, but it lacks in flexibility when data and voice integration is required. Evaluation of the improvements on the spectral efficiency that PRMA can offer to TDMA systems can be found in [35, 36].

2.5 CDMA Protocols

In CDMA protocols the entire transmission bandwidth is shared between all users all the time [2, 37], and the multiple access property is achieved by assigning each user a different code. This code is used to transform a user's signal in a wideband signal (spread spectrum signal). Upon reception of multiple wideband signals, the receiver uses the code assigned to a specific user to transform the wideband signal back to the original signal. During this process, the desired signal power is compressed into the original signal bandwidth while the wideband signals of the users remain wideband signals and appear as noise when compared to the desired signal. As long as the number of interfering users is not too large, the signal-to-noise ratio will be large enough to extract the desired signal without error. Thus, in this case the protocol behaves as a contentionless protocol. However, if the number of users rises above a certain limit, the interference becomes too large for the desired signal to be extracted and contention occurs, thus making the protocol interference limited.

A typical way the wideband channel is shared among the users is the *direct sequence CDMA* (DS-CDMA), which will be described in more detail in the following sections. Since CDMA, especially in its direct-sequence version, is one of the most important multiple access technique for future wireless communication systems, a paragraph will be provided with more details on a specific topic such as power control in CDMA systems. Furthermore, a comparison between TDMA and CDMA is provided and the opportunity to combine them is discussed in Section 2.5.2 where examples of hybrid TDMA/CDMA systems are also presented.

2.5.1 DS-CDMA

DS-CDMA is obtained by spreading the user signals onto a bandwidth much larger than an individual user's information rate. A typical way the wideband channel is shared among the users is the direct sequence CDMA, where each user's information symbols are spread over the wideband channel by its unique *signature sequence* $(c_1, c_2, \ldots, c_N)$ characterized by a chip waveform, $p_{T_c}(t)$:

$$\int_{-\infty}^{\infty} p_{T_c}(t) p_{T_c}(t - nT_c)dt = 0 \quad \forall n, n \neq 0 \tag{2.11}$$

where T_c is the chip period.

The number of chips per bit, N, is known as the *spreading factor*. The overall spread of the bandwidth is called *processing gain G*, defined as the ratio between the bandwidth of the signal transmitted and the bandwidth of the information signal. Denoting with $b_i^{(j)} \in \{\pm 1\}$ the binary data stream that the user i intends to transmit (or, the base station intends to transmit towards the user i), the transmitted signal by the user (or towards the user) can be written:

$$s_i(t) = \sqrt{P_0 \eta_i} \, b_i(t) a_i(t) \tag{2.12}$$

where

$$a_i(t) = \sum_{j=1}^{G} (-1)^{c_{ij}} p_{T_c}(t - (j-1)T_c) \tag{2.13}$$

is the spreading sequence assigned to the user i,

$$b_i(t) = \sum_{j=-\infty}^{\infty} b_i^{(j)} g_{T_b}(t - (j-1)T_b) \tag{2.14}$$

$g_{T_b}(t)$ is the impulse response of the transmit filter and $1/T_b$ is the bit rate; $P_0 \eta_i$ is the transmission power by/to user i: it is written as a product of a nominal value P_0 equal for each user and a factor η_i that depends on the power control strategy adopted. This notation for the transmission power is useful as a base station-mobile link analysis is considered hereinafter where the focus is on the transmission power level rather than on the received power levels. In the case of a general asynchronous system with K users, the signal received by the user of interest or the base station can be written as

$$r(t) = \sum_{j=1}^{K} \sqrt{P_0 \eta_i} R_j b_j \left(t - \tau_j\right) a_j \left(t - \tau_j\right) \cos\left(2\pi f_c t + \theta_j\right) = n(t)$$

(2.15)

where f_c is the frequency carrier, τ_j and θ_j represent a random delay and a random frequency phase for the jth user. Without lack of generality it has been assumed that the random delay and frequency phase of the user of interest are zero. $n(t)$ denotes the thermal noise, modeled as an additive white Gaussian noise with two-sided power spectral density $N_0 / 2$. The R_j's represent the channel attenuation experienced by each signal. In the case of the mobile-to-BS link (uplink or reverse link), the channel impairments experienced by each channel are different. In the opposite link (downlink or forward link), desired signal and interference fade simultaneously ($R_i = R_j \forall i, j$). To draw some of the key aspects of CDMA systems, the reverse link will be considered in the rest of this section (unless explicitly mentioned).

The receiver has the task of discriminating among users. In what follows, the focus is on the class of *linear receivers* (i.e., receivers that operate linearly on the total received signal to demodulate the symbol of a particular user).

Matched Filter

The simplest linear receiver is the one that is matched to the signature sequence of the desired user. This matched filter receiver is the receiver used in the IS-95 standard. The operation of a conventional DS demodulator may be regarded as computation of the correlation of the received signal with a specific user signature sequence, followed by a hard decision on the correlator output. Let us focus on the demodulation of user i's symbols and consider the reception of the data bit $b_{i,0}$ transmitted by the user i in the instant 0. The output Z_i of the correlation receiver is

$$Z_i = \sqrt{\frac{P_i \eta_i}{2}} R_i T_b b_{i,0} + N_i + I_i$$

(2.16)

where N_i is a zero mean Gaussian random variable of variance $N_0 T_b/4$ and I_i represents the contribution of the multiple access interference (MAI) whose expression is found to be [2, 37]

$$I_i = \sum_{k=2}^{K} \sqrt{\frac{P_0 \eta_k}{2}} R_k \cos(\psi_{ik}) B(i, k, \tau_{ik}) \qquad (2.17)$$

where $\psi_{ik} = \theta_{ik} - 2\pi f_c \tau_{ik}$ and

$$B(i, k, \tau) = \int_0^{T_b} b_i(t - \tau) a_i(t - \tau_i) a_k(t) dt \qquad (2.18)$$

depends on the cross-correlation properties of the spreading codes.

When all user signals are generated or may be coordinated within a single site, it makes sense to adopt synchronous CDMA (S-CDMA), whereby all signals share the same chip and symbol references. The satellite-to mobile link is well suited for S-CDMA. The S-CDMA can be achieved by (2.15), letting $\tau_j = \tau \forall j$. In that case, sequences can be chosen to be orthogonal to each other in order to avoid interference. Families of *orthogonal codes* like Walsh-Hadamard (WH) binary functions (Orthogonal-CDMA) or the so-called *quasi-orthogonal codes* such as the preferentially-phased Gold codes [38–40] can be used. Orthogonal-CDMA systems are hard-limited in capacity by the maximum number of available orthogonal codes. Moreover, multipath can cause an orthogonality loss since several delayed replicas of the signal are superimposed together at the receiver.

CDMA systems are usually *asynchronous* in the uplink, which means that there is a random relative delay between users so that a symbol of a user overlaps with two partial symbols of an interferer. With asynchronous CDMA (A-CDMA), it is customary to assign each user a different random-like pseudo-noise (PN) sequence characterized by low cross-correlation properties.

Looking at the general problem of demodulation, which is associated with extracting good estimates of each user (coded) symbol and soft decisions to be used by the channel decoder [38], a relevant performance measure is the signal-to-interference ratio (SIR) of the estimates, which can be taken as a quality-of-service measure for the user. Strictly speaking, the SIR does not completely characterize performance such as bit error probability, since the interference from other users is not necessarily Gaussian. However, it has been found in practice to be a reasonable measure, and its use is further justified rigorously in [41–43] for a large system with many interferers, using the Central Limit Theorem. Let us focus on the more general case of A-CDMA. Let us assume that the carrier phase, time delay, and data symbols of any user are independent random variables and independence is also assumed for different users. Moreover, independent random binary sequences are taken as spreading sequences. PN sequences are a very close approximation to true random sequences for which the model is appropriate. MAI is modeled as a Gaussian random variable with zero mean and variance [41]:

$$\mathrm{var}[I_i] = \frac{T_b^2}{3G}\sum_{k=2}^{K} P_0 \eta_k E[R_k^2]$$ (2.19)

where T_b is the bit interval and $E[\cdot]$ denotes the statistical mean of the random variable.

The average SIR is

$$SIR = \frac{\sqrt{\dfrac{P_0\eta_i}{2}}\,T_b R_i}{\sqrt{\dfrac{N_0 T_b}{4} + \mathrm{var}[I_i]}} = \left[\frac{E_{ib}}{2N_0} + \frac{E[R_k^2]}{3G}\sum_{k=2}^{K}\frac{\eta_k}{\eta_i}\right]^{-1/2}$$ (2.20)

where $E_{ib} = P_0\eta_i T_b$ is the bit energy for the user i.

From the expression of the MAI, it is easily seen that a user can completely dominate the others and draw out the signals of the other users. This is known as the *near-far* problem, because it is likely to occur, without power control, when an interferer is near to the desired signal receiver, and the user is far. Equalization of the user-received powers is a simple power-control strategy to counteract this effect. The opposite link does not present this problem, since received signals come from the same transmitter and undergo the same channel conditions. Nonetheless, a proper power control is still fundamental to reduce the intercell interference and energy consumption. In particular, the reduction of the energy consumption at the base-station is an important issue in satellite applications.

Rake Receiver

Due to multipath propagation, the signal energy (pertaining, for example, to a single chip of a CDMA waveform) may arrive at the receiver across clearly distinguishable time instants. For certain time delay positions there are usually many paths nearly equal in length along which the radio signal travels. For example, path with a length difference of half of a wavelength (at 2 GHz this is approximately 7 cm) arrive at virtually the same instant if compared to the duration of a single chip, which is 78m at 3.84 Mcps. As a result, signal cancellation, called fast fading, takes place as the receiver moves across even short distances. Signal cancellation is the summation of several weighed phasors that describe the phase shift (usually modulo radio wavelength) and attenuation along a certain time instant. In principle, the attenuation of the received signal power can reach an infinite value in dB when phase cancellation of multipath reception occurs.

Because of the underlying geometry causing the fading and dispersion phenomena, signal variations due to fast fading occur several orders of magnitude more frequently than changes in the average multipath delay profile.

Statistics of the received signal energy for a short-term average are usually well described by the Rayleigh distribution [44].

Rake receivers in CDMA systems provide multipath diversity against fading. A Rake receiver identifies the time delay positions at which significant energy arrives and allocates correlation receivers to those peaks. A matched filter is used for determining and updating the current multipath delay profile of the channel. This measured and possibly averaged multipath delay profile is then used to assign the Rake fingers to the largest peaks. Then, within each correlation receiver, by estimating the momentary channel state (value of the weighed phasor) for a particular finger, the received signal is rotated back, so to undo the phase rotation caused by the channel. Such channel-compensated symbols can be simply summed together to recover the energy across all delay positions. This processing is also called *maximum ratio combining.*

Note that multiple receive antennas can be accommodated in the same way as multiple paths received from a single antenna: by just adding additional Rake fingers to the antennas, it is possibly to receive all the energy from multiple paths and antennas. This is another form of diversity (macrodiversity) with respect to the diversity related to the paths received by a single antenna (microdiversity). From the Rake receiver perspective, there is essentially no difference between these two forms of diversity reception.

Multiuser Detection

In a conventional receiver, power control is the only counter-measure available for the near-far problem. The near-far problem is not intrinsic in a direct sequence CDMA system, but it is due to the suboptimality of the matched filter receiver that is optimized for an AWGN channel and does not take into account the structured nature of the interference [45]. Performance gain can be achieved with receivers that take into account the real structure of the *multiuser* interference. In multiuser detection, code and timing information of multiple users are jointly used to better detect each individual user. The first multiuser receiver proposed by Verdu has the property of minimizing the probability of symbol detection error [46]. However, its complexity grows exponentially with the number of users in the system. Linear multiuser receivers were later proposed which have lower complexity and retain much of the performance advantage over the conventional matched filter receiver. The subject of the multiuser detection (MUD) in CDMA system is a hectic research area [47–53]. Even if MUD receivers have been conceived to avoid the exploitation of power control to counteract the near-far problem, power control is still a fundamental resource allocation means [54].

Power Control

Since spread spectrum systems do not explicitly schedule time or frequency slots among users, one of the central mechanisms for the interference management is *power control.* Each user varies its access to resources by adapting its transmit power to the changing channel and interference conditions. When the link mobile-to-base station (reverse link) and a traditional correlator receiver are considered, power control is needed for facing the near-far problem. In the common power control policy, analyzed in many references ([31, 54–56]), users equalize their received powers so that nearby users do not dominate over the far-away users. An implicit assumption is that the system under study is for one class of service (e.g., voice). In the reverse link of a CDMA system with multiple classes of services characterized by different bit rates and QoS requirements, the optimum power control policy envisages a target power level assigned to each user, which depends on the source rate and the QoS requirement [57]. When the base station-mobile station link (forward link) is considered, the near-far problem is absent and a proper power control aims at optimizing (minimizing) the exploitation of the power available at the base station. If this aspect is of secondary importance in *interference limited* systems like cellular systems, it is fundamental in systems with constraints on the transmission power at the base station. Therefore, it is fundamental in satellite systems where the available onboard power is limited. Furthermore, a proper power control strategy, seen as a central mechanism of resource management in CDMA systems, is also needed when multiuser receivers are considered [54]. Power control strategies are implemented through the combination of two different techniques:

1. *Open-loop power control:* to compensate the path attenuation and the slow shadowing. Such a control is based on an estimate of the deviation of the received pilot power with respect to its nominal level. This estimate is used to correct the uplink power, assuming symmetry in the loss between downlink and uplink.

2. *Closed-loop power control:* to provide fine correction of signal level variations and to compensate for MAI-induced impairments. A quasi-real-time estimation of some parameter used to measure the quality of service at the receiver is performed at the other end of the link. According to this estimation, incremental adjustments of the transmitted power are then sent back by the receiver in the form of inband signaling packets.

The closed loop power control may consist of an inner loop and outer loop. A functional scheme for a closed-loop power control scheme in a base station-to-mobile link is shown in Figure 2.10.

The inner loop control is fast acting and consists of a SIR-based closed-loop power control. In each frame it measures E_b / N_T (i.e., the energy per chip divided by the total noise density, where total noise includes both thermal and

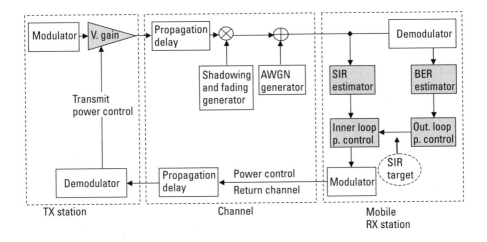

Figure 2.10 Power control scheme.

interference). It compares this measurement against a target value and commands the transmitter to increase or decrease the power accordingly. Actually, this threshold decision alone does not determine whether to toggle the power control bit up or down. The algorithm controlling the bit also takes into account the delay and the most recent bits that have been transmitted. This is because it takes a number of frames before the user can see any effect from previous power control bits. If the power control bits are toggled up and down without taking this effect into account, the response would be highly oscillatory. Over this inner loop control is a slower acting outer loop. The job of the outer loop is to set or control the target value of the E_b / N_T used by the inner loop. The outer loop can set this to the proper value by observing which frames are received in error. Its goal is to adjust the target value so that the frame error rate is neither too good nor too bad. The outer loop process is required whenever the channel state statistic changes and, hence, the link performance can no longer be controlled through the observation of a pure signal-to-noise measure.

2.5.2 TDMA and CDMA

In FDMA/TDMA-based systems the capacity is hard-limited (e.g., a noticeable performance degradation is observed when the number of users is higher than a maximum value depending on the system design). Therefore, the system must be designed for worst-case interference characteristics, not providing an efficient use of power and spectrum resources in systems where the number of potential users is much greater than the number of simultaneous users at a given time. Power control, DCA, and all the other techniques described in Section 2.5.1, can enhance the capacity by providing a more dynamic and flexible resource allocation. The main advantages of CDMA come from its interference-averaging capability. The capacity is limited by the total interference power. The number of interferers is irrelevant, as is the actual power of each user. The interference is the sum of a large number of interfering signals that fade independently. The law of large numbers ensures that the system can be designed for average interference characteristics, rather than for the worst case. The fluctuation in the total interference power is much less than the power fluctuation of a single interfering signal. Hence, the fading margin of a CDMA system can be significantly smaller than the margin for a FDMA/TDMA system. Furthermore, exploiting voice activity detection and sectorization can effectively reduce the total interference power level. In [58], it is shown that for voice traffic, these aspects largely determine the capacity gain of CDMA systems with respect to FDMA/TDMA systems. However, tight power control requirements are required if correlation receivers are used. MUD receivers can reduce the requirements on the power control, but a properly designed CDMA system with MUD receivers does not exploit all the benefits coming from the interference averaging-effect since the system turns out to have a hard-limited capacity as in the case of FDMA and TDMA systems [59].

On the other hand, when asymmetries in the traffic or channel characteristics can be foreseen, the fixed allocation achieved in TDMA systems could be more efficient in the resource exploitation with respect to CDMA systems. Actually, as will be outlined in Chapter 6, effective resource allocation techniques in CDMA systems somehow induce a time division in the system. This observation can help in understanding the trade-off and advantages that can be achieved by combining TDMA and CDMA. In the next section, some examples of hybrid TDMA/CDMA systems are described.

2.5.2.1   Spread ALOHA

In [11], a simple linear transformation of the ALOHA multiple access—indicated as spread ALOHA, a combination of CDMA and TDMA—is proposed. Each user adopts the same spreading code. In the upper part of Figure 2.11 it is shown what happens if two users access the medium at the same time in case the standard ALOHA access is used.

In the lower part of the same figure, the two spread signals are shown in the time domain. Even if the spreading code is the same, if the chip phase is slightly different, the separation of the two signals is still possible when one receiver is synchronized to one sender and another one to the other sender. The spreading code should have good autocorrelation for the synchronization process and correlation should be low if the phase differs slightly. The maximum throughput is about 18%, which is very similar to ALOHA, but in addition the approach benefits from the performance of spread spectrum techniques in fading channels.

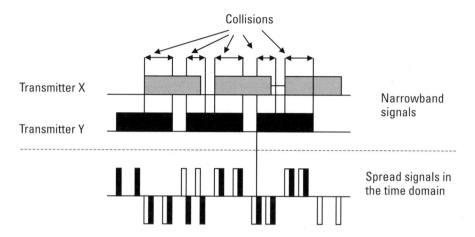

Figure 2.11 Spread spectrum ALOHA principle.

2.5.2.2 Time Division–Code Division Multiple Access

Time division–code division multiple access (TD-CDMA) systems combine the characteristics of CDMA and TDMA multiplexing techniques. As in a TDMA technique, the transmission on each frequency bearer is organized in frames divided in time slots. As a difference with respect to TDMA technique, each time slot can be shared by more than one user by using a CDMA approach.

The TD-CDMA technique is utilized in the TDD mode of the UTRA proposal. The design goals in terms of maximum data rates achievable in the TDD mode are at least 384 Kbps for bidirectional transmission in microcells and 2 Mbps for unidirectional transmission in picocells. The TD-CDMA concept provides the performance as well as the flexibility to fulfill these requirements. Figure 2.12 shows a frame structure for UTRA-TDD.

Each time slot comprises a maximum of 16 orthogonal spreading codes. Pooling of CDMA codes and TDMA time slots can easily provide the capacity demanded by a certain user (i.e., by the allocation of several codes or time slots or both to a single user). Up to eight users can share up to 16 orthogonal spreading codes. Due to the small number of spreading codes per time slot, this technique enables the multiuser detection or joint detection to mitigate the intracell interference. Therefore, the requirements on power control accuracy are relaxed. Furthermore, each time slot can be allocated to either the uplink or the downlink. In one frame, the switching point between the uplink and downlink allocation of time slots can change once (single switching point configuration) or several times (multiple switching point configuration) per frame (Figure 2.13).

In any configuration, at least one time slot has to be allocated to the uplink and at least one to the downlink. By just changing the allocation of the time slots within the frame between the uplink and downlink, higher spectrum

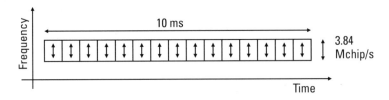

Figure 2.12 Frame structure for UTRA-TDD.

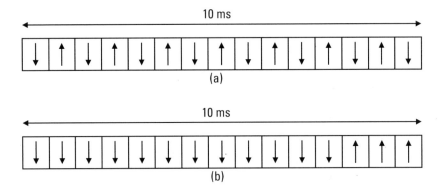

Figure 2.13 Switching point configurations: (a) multiple switching point configuration; and (b) single switching point configuration.

efficiency is achieved since the need for higher overall data rate in one link can be satisfied by taking part of releasable resources of the other link. This is the major advantage of TDD mode.

Channel Allocation

In UTRA-TDD mode, a physical channel, also indicated as resource unit (RU), is characterized by a combination of its carrier frequency, time slot, and spreading code. Due to the TDMA component, interference avoidance algorithms by means of DCA can be applied to both coordinated and uncoordinated operation. The resource allocation to cells is called slow dynamic channel allocation, and the allocation of resource units to bearer services is called *fast* DCA [60–62].

Joint Detection

Data detection has to face two problems at the same time: ISI between consecutive symbols associated with a single user due to multipath propagation; and MAI between data symbols of different users caused by the lack of orthogonality of the spreading codes at the receiver. Joint detection receivers are multiuser detectors, which combat both ISI and MAI by exploiting the knowledge about spreading sequences and channel impulse responses. Detailed information on joint detection techniques and their applications can be found in [60, 63–69]. These techniques result in a very large system of equations and the algorithms that do not exploit the structural properties of the matrices have computational complexity of $O(M^3)$, where M denotes the product of the number of active users and the number of symbols per data block. One of the main objectives of the research is to reduce the computational complexity of these algorithms [68, 69].

2.5.3 Multicarrier CDMA

The use of conventional CDMA does not seem to be realistic when the data transmission rate is in the order of a hundred Mbps, due to the severe intercode interference and the difficulty to synchronize such a fast sequence. Techniques of reducing both the symbol rate and the chip rate are essential in this case. A multicarrier transmission technique, orthogonal frequency division multiplexing (OFDM), combined with CDMA has been proposed to solve this problem. Such a system that combines orthogonal multicarrier modulation and CDMA is usually referred to as multicarrier CDMA or OFDM-CDMA and, in recent years, has drawn a wide interest in wireless personal multimedia communications [3, 70–73]. Various multicarrier CDMA schemes have been proposed [74–78]. The common point of these proposals is to change the conventional serial transmission of data/chip stream into parallel transmission of data/chip symbols over a large number of narrowband orthogonal carriers, hence the bit and chip duration is increased proportionally. Multicarrier CDMA (MC-CDMA) schemes can be categorized in two groups [70]:

1. Spreading in the frequency domain;
2. Spreading in the time domain similar to a conventional DS-CDMA scheme.

A general introduction to OFDM will first be provided before introducing MC-CDMA schemes.

2.5.3.1 OFDM

OFDM is a multicarrier transmission technique [79]. Figure 2.14 shows the general structure of a multicarrier system.

The original data stream of rate R is multiplexed into N parallel data stream at rate $R_{MC} = 1/T_{MC} = R/N$. Each of the data streams is modulated with a different frequency, and the resulting signals are transmitted simultaneously in the same band. The receiver consists of N parallel receiver paths. If $R = 8$

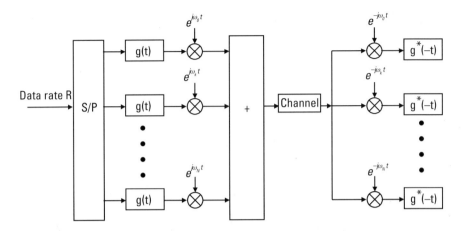

Figure 2.14 Multicarrier approach.

Msymbol/s and the channel delay spread is $\tau_{max} = 2\,\mu$sec, the single carrier system is characterized by an ISI of $\dfrac{\tau_{max}}{T} = 16$ symbols.

Due to the increased distance between transmitted symbols, the ISI for each subsystem in the multicarrier system reduces to

$$\frac{\tau_{max}}{T_{MC}} = \frac{\tau_{max}}{NT} \tag{2.21}$$

In an OFDM system, the subcarriers are made mathematically orthogonal to each other, preventing interference between closely spaced carriers. This allows them to be spaced very closely with no overhead as in FDMA. The orthogonality of the carriers means that each carrier has an integer number of cycles over a symbol period, as shown in Figure 2.15.

The receiver acts as a bank of demodulators where the downconverted signal of each carrier is integrated over a symbol period to recover the raw data. If each subcarrier, in the time domain, has exactly an integer number of cycles in the interval T, then the integration process results in zero contribution from all these subcarriers.

In 1971, Weinstein and Ebert applied the discrete Fourier transform (DFT) to parallel data transmission systems as a part of the modulation and demodulation process [80]. Using DFT-based multicarrier techniques allows for the elimination of the banks of subcarrier oscillators and coherent demodulators required by frequency division multiplex, and a completely digital implementation could be built around special-purpose hardware performing the fast Fourier transform (FFT), an efficient implementation of the DFT. Recent advances in

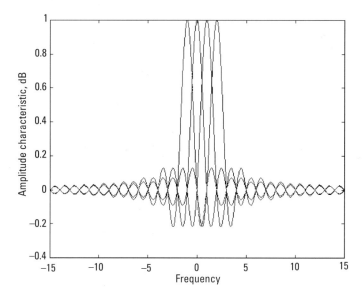

Figure 2.15 Spectra of four orthogonal carriers.

very large scale integration (VLSI) technology render high-speed, large-size, FFT chips commercially affordable. By using this approach, both transmitter and receiver are implemented using efficient FFT techniques, which reduces the number of operations from N^2 in DFT down to $N\log N$ [80]. Therefore, OFDM provides an efficient way to deal with multipath if the amount of ISI in each subcarrier does not bring to the need for an equalizer in the receiver, hence reducing the receiver implementation complexity. OFDM itself does not get a diversity gain in presence of frequency selective channels. The information carried by subcarriers subjected to deep fade is lost. To get a higher robustness against frequency selective channels and narrowband interference, OFDM is usually combined with interleaving and soft decision channel decoding. This combination is called coded-OFDM (COFDM) and it has been chosen for the Digital Audio Broadcasting (DAB) and terrestrial Digital Video Broadcasting (DVB-T) in Europe [81]. So far, digital broadcasting has been the main application for OFDM schemes. OFDM has been proposed for digital cellular systems in the mid-1980s [82]. Severe disadvantages such as difficulty in subcarrier synchronization, sensitivity to frequency offset, and nonlinear amplification have made OFDM not suitable for outdoor mobile systems; it still represents, however, a viable modulation type for high-data rate indoor wireless communications [83]. In fact, it has been incorporated into standard by the ETSI HiperLan2. The IEEE 802.11 standards group has chosen OFDM modulation for wireless LAN operating at bit rates up to 54 Mbps at 5 GHz. Recently, OFDM modulation combined with dynamic packet assignment with wideband

5-MHz channels is proposed for high-speed packet data wireless access in macrocellular and microcellular environments, supporting a family of peak bit rates ranging from 2 to 10 Mbps [84].

2.5.3.2 Spreading in the Frequency Domain: MC-CDMA

The original data stream is spread by using a given spreading code and then different subcarriers are modulated with each chip. A fraction of the symbol corresponding to a chip of the spreading code is transmitted through a different subcarrier. There is only one scheme in this category and it is referred to as MC-CDMA. In Figure 2.16, an MC-CDMA transmitter scheme is shown.

For MC transmission, it is essential to have frequency nonselective fading over each subcarrier. Therefore, if the original symbol rate is high enough to undergo frequency-selective fading, the signal needs to be first serial-to-parallel converted before spreading over the frequency domain.

Therefore, the original data stream is first converted into P parallel streams $(a_{j,0}, a_{j,1}, \ldots, a_{j,P-1})$. L denotes the total number of chips per data bit or the processing gain. The system transmits L chips of a data symbol in parallel on different

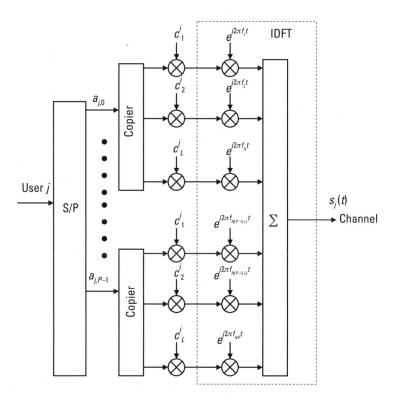

Figure 2.16 Spreading in the frequency domain: MC-CDMA transmitter.

carriers, one chip per carrier. Thus, the chip duration T_s of MC-CDMA system is the same as the bit duration T_b. The total number of subcarriers per original data stream is $N_c = P\,L$. The serial-to-parallel conversion of the original data stream may be needed to increase the chip duration to avoid frequency-selective fading. In order to achieve frequency diversity, the assignment of the N_c carriers to the N_c chips is carried out to maximize the frequency separation among carriers conveying the chips of the same data bit. Hence the L data chips in the pth data stream ($p = 1, 2, ..., P$) are transmitted on the L carriers with frequencies $\{f_{p+(l-1)P}, l = 1, 2, ..., L\}$ and the adjacent frequency separation between these carriers is P/T_s. A MC-CDMA receiver requires coherent detection for successful despreading operation. Figure 2.17 shows the MC-CDMA receiver for the jth user.

After downconversion, the mth subcarrier corresponding to the received data $a_j(i)$ is first coherently detected with DFT and then multiplied by the gain q_m^j to combine the received signal energy scattered in the frequency domain. The decision variable is the sum of the weighed baseband components and it is given by

$$d^j(t = iT_s) = \sum_{m=1}^{L} q_m^j y_m \quad \text{with} \quad y_m = \sum_{j=1}^{J} z_m^j(iT_s)a_j(i)c_m^j + n_m(iT_s)$$

(2.22)

where y_m and $n_m(iT_s)$ are the complex baseband component of the received signal after downconversion with subcarrier frequency synchronization and the complex Gaussian noise at the mth subcarrier at $t = iT_s$, respectively; $z_m^j(iT_s)$ is the

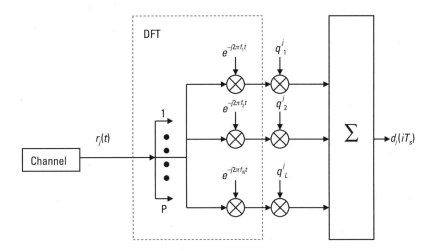

Figure 2.17 MC-CDMA receiver scheme.

complex envelope of the mth subcarrier for the user jth, and J is the total number of active users. Different detection strategies are described in Table 2.1.

Each detection strategy is obtained by different choices for q_m^j.

The minimum mean square error combining (MMSEC)-based MC-CDMA is a promising scheme in a downlink channel, although estimation of the noise power as well as the subcarrier values is required. MUD strategies that require the knowledge of the spreading codes assigned to all the active users are applicable in the downlink but they are needed in the uplink application since the code orthogonality among users is totally distorted by the channel frequency selectivity.

2.5.3.3 Spreading in the Time Domain: MC-DS-CDMA and MT-CDMA

The serial-to-parallel converted data streams are spread by using a given spreading code, and then different subcarriers are modulated with each data stream. There

Table 2.1
Detection Strategies for MC-CDMA Systems

Detection Strategy q_m^j	Characteristics
Orthogonal restoring combining (ORC) $$q_m^j = c_m^j \left(z_m^j \right)^* / \left| z_m^j \right|^2$$	It can eliminate multiuser interference perfectly
	Low-level subcarriers tend to be multiplied by high gains, and the noise components are amplified at weaker subcarriers
	It is applicable only for the downlink
Controlled equalization (CE) $$q_m^j = c_m^j \left(z_m^j \right)^* / \left| z_m^j \right|^2$$	Decision is made based on the sum of baseband components of subcarriers whose amplitudes are larger than a detection threshold thus suppressing the noise amplification effect in ORC
Equal gain combining $$q_m^j = c_m^j \left(z_m^j \right)^* / \left| z_m^j \right|^2$$	
Maximal ratio combining $$q_m^j = c_m^j \left(z_m^j \right)^*$$	In case of one user, it minimizes the BER
Minimum mean square error combining (MMSEC) $$q_m^j = c_m^j \left(z_m^j \right)^* / \left(\sum_{i=1}^{J} \left| z_m^j \right|^2 + N_0 \right)$$	Downlink
	For small $\left| z_m^j \right|$ the gain becomes small to avoid excessive noise amplification
	For large $\left| z_m^j \right|$ it becomes proportional to the inverse of the subcarrier envelope $\left(z_m^j \right)^* / \left| z_m^j \right|^2$ to recover orthogonality among users

are two schemes in the second category: multicarrier direct-sequence CDMA (MC-DS-CDMA) and multitone CDMA (MT-CDMA). In Figure 2.18 the transmitter schemes of these two techniques and the related power spectrum of the transmitted signal are shown.

MC-DS-CDMA transmitter spreads the S/P converted data stream by using a given spreading code in the time domain so that the resulting spectrum of each subcarrier can satisfy the orthogonality condition with the minimum frequency separation $1/T_s$. As in any MC transmission scheme, it is crucial to have nonselective fading over each subcarrier. The receiver is usually composed of N normal coherent (non-Rake) receivers, as shown in Figure 2.19(a).

Therefore, with no forward error correction among subcarriers, this scheme is unable to achieve any frequency diversity gain. In [85], a MC-DS-CDMA scheme is proposed with a larger subcarrier separation in order to yield both frequency diversity improvement and narrowband interference suppression.

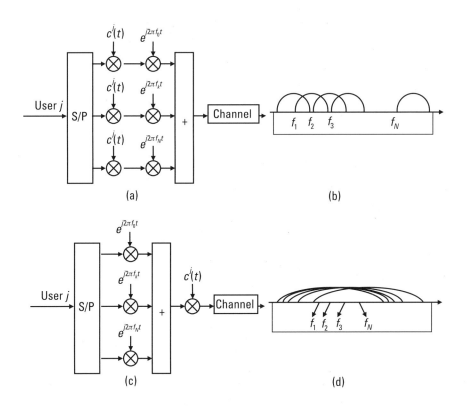

Figure 2.18 Spreading in the time domain: (a) MC-DS-CDMA transmitter; (b) power spectrum of the MC-DS-CDMA transmitted signal; (c) MT-CDMA transmitter; and (d) power spectrum of the MT-CDMA transmitted signal.

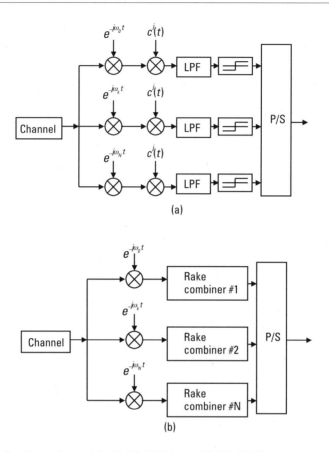

Figure 2.19 Receiver scheme: (a) MC-DS-CDMA; and (b) MT-CDMA.

The MT-CDMA transmitter spreads the S/P converted data stream using a given spreading code in the time domain so that the spectrum of each subcarrier prior to spreading operation can satisfy the orthogonality condition with the minimum frequency separation $1/T_b$. Therefore, the resulting spectrum of each subcarrier no longer satisfies the orthogonality condition. This scheme suffers from intersubcarrier interference, but it can use longer spreading codes in proportion to the number of subcarriers if compared with a DS-CDMA system. This results in a reduction of ISI and MAI. In channels where this improvement is dominant, the MT-CDMA scheme can outperform the DS-CDMA scheme. In Figure 2.19(b) the optimum receiver for the additive white Gaussian channel is shown. It is composed of N Rake combiners. Furthermore, a decision feedback equalizer (DFE), a linear equalizer (LE), and a linear joint multiple access interference canceler/equalizer (JEIC) suited for frequency-selective fast fading channels have been proposed, all of which have multiple input–multiple output (MIMO) type structures based on the MMSE criterion.

2.6 MAPs for Wireless Multimedia Communications

Traditional voice-based multiple access protocols would perform poorly in future wireless networks. For instance, even if a simple implicit resource reservation, such as in PRMA protocol, requires simpler acknowledgment, this approach is no longer sufficient for high bit rate services, which may require the allocation of multiple slots in a frame to a single user. In this case, the BS should respond by specifying explicitly the reserved set of resources. Portable wireless devices have severe constraints on size, energy consumption, and bandwidth. The energy consumption needed for communication and computation will limit the functionality of mobiles, and so it becomes fundamental to increase the energy efficiency [86]. For the wireless channel with constraints on the admissible delay, delay jitters, and average transmission power, the use of contention for the reservation and of packet retransmission for solving the contention become key issues. Furthermore, a single terminal should be able to access different services simultaneously. Therefore, it should be possible to handle simultaneously different traffic classes, with different QoS requirements (BER, delay) and varying bit rates in addition to supporting bursty traffic sources and asymmetric communications applications, such as Internet downloads [87–89]. In this context, the introduction of more or less complex and dynamic mechanisms to assign priority at MAC layer turns out to be necessary.

The throughput as previously defined is clearly no longer a proper comparison criteria among different protocols and solutions. Important issues to be accounted for the design of MAC protocols are as follows:

- QoS support;
- Priority access support;
- Complexity;
- Energy efficiency.

In the following sections, protocols are presented that have been designed keeping in mind some of the above requirements. An overview of these protocols is shown in Table 2.2.

The main features of each protocol together with the scenario for which it has been designed/optimized is now discussed.

2.6.1 Dynamic Packet Reservation Multiple Access

Dynamic packet reservation multiple access (DPRMA) has been developed for ATM wireless networks [90]. With respect to the PRMA protocol, DPRMA includes the following additional elements: (1) priority, and (2) the possibility of assigning users more than one slot per frame according to the amount of required bandwidth. The user submits an initial request, or a change in rate

Table 2.2
Some MA Protocols for Multimedia Traffic in Wireless Networks

MA Protocol	Specific Environment	QoS Support	Priority Access	Other Properties
Dynamic-PRMA	Wireless ATM networks	Based on required BW	Real-time versus non-real-time	Full-sized request slots for contention periods
Centralized-PRMA	Microcellular environment	Based on loss, delay, traffic requirements	Scheduling of polling sequence	Mini-sized request slots, central role of BS
Multidimensional PRMA	UTRA/TDD	Based on traffic rate and delay	Different TX probabilities	No multi-subslots allocation in the same frame
WCDMA MAC	WDMA standard	Based on required BER and delay	Different TX format specified	The stability of the protocol depends on higher layer functionalities (CAC, CC)

request, by setting the appropriate reservation request (RR) bit in the header of the uplink slot. Via several reservation acknowledgments (RA) bits in the header of downlink messages, the base station communicates the results of the contention period and allocates as much of the user requested rate as possible. For time-dependent traffic, as many of the available empty slots as necessary are used to fill the user request. When additional slots are needed, the slots occupied by data traffic are preempted and the data users are placed in a queue to wait for further service when the bandwidth becomes available again. If the guaranteed rate of real-time users still cannot be met, packets are dropped. Data traffic is lost only when the buffer for preempted packets overflows.

All users can submit reservations, but real-time requests have a higher priority than non-real-time traffic.

DPRMA is characterized by a simple approach to the priority access for different traffic classes (i.e., non-real-time versus real-time) and by a simple bandwidth assignment mechanism. The major disadvantage of the protocol is the bandwidth wasting, as requests in the contention periods are made in full-size slots [90].

2.6.2   Centralized-PRMA

Centralized-PRMA (CPRMA) is a packet-switched multiple access protocol, especially suitable for a microcellular environment [91]. It adopts some basic concepts of PRMA, but presents more centralized functions, thus exploiting the inherently centralized nature of a cellular system. With respect to other previously described protocols, a more complex scheduling algorithm is implemented, which makes the system more flexible and, hence, more efficient in supporting different services. Two separate time-slotted channels are used for communication within each cell: the uplink channel, which conveys the information from the mobile stations to the base station; and the downlink channel, which is exclusively used by the base station. As is shown in Figure 2.20, in the available slots of the uplink reservation requests are transmitted in minipackets containing such information as terminal identifier, time spent by the first packet in the buffer, cyclic redundancy check (CRC) field.

This information is used by a scheduling algorithm at the base station to generate the polling sequence for the reserved terminal. The polling command specifies whether a slot of the uplink channel is available or reserved, identifying the mobile station enabled to transmit. Therefore, the polling mechanism provides the coordination in the reserved slots among the active mobile stations. When the base station issues the polling command with the remote terminal

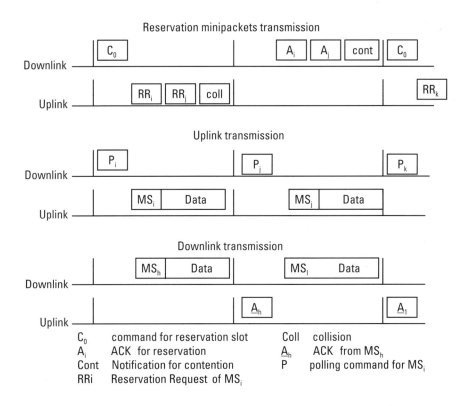

Figure 2.20 Data and signaling channels of the CPRMA protocol.

identifier, the terminal can transmit packets on the assigned reserved slots and it keeps on transmitting as long as it remains active. If a collision occurs, the MS remains in a *contending state* where a cycle for collision resolution is initiated, continuing until all of collided minipackets have successfully been transmitted. CPRMA, at the price of a more complex approach to the priority assignment with respect to the DPRMA, provides a more accurate statistical combination of traffic, better efficiency, and higher performance in presence of multimedia traffic.

2.6.3 Multidimensional PRMA with Prioritized Bayesian Broadcast

Multidimensional-PRMA (MD-PRMA) is an adaptation of the PRMA to any hybrid time division multiple access schemes such as FDMA/TDMA and CDMA/TDMA and multimedia traffic [92]. In MD-PRMA, slots are defined not only in the time domain, but also in an additional dimension, either the frequency or the code domain. This protocol has been proposed for UTRA TDD [93].

Time slots are subdivided into Q subslots or simply slots, either using an FDMA component or a CDMA component by introducing a code. The subdivision in frames and slots in case of hybrid TDMA/CDMA and $Q = 8$ is shown in Figure 2.21.

On the downlink, the base station indicates, slot by slot, if the slot is in a contention phase (*C*-slot) or if it is reserved for the information transfer (*I*-slot).

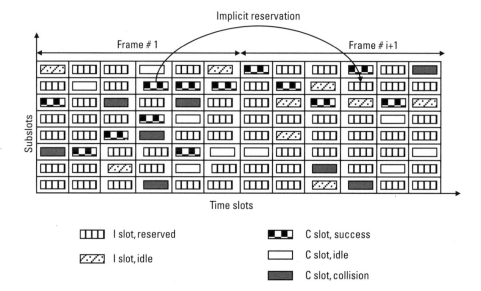

Figure 2.21 Slots and frames in MD-PRMA.

A mobile that has no reservation but is admitted to the system may only access slots in contention mode (C-slots), with some service and time slot–specific permission probability p_x. Probabilities for each type of service and each time slot of the next uplink frame are broadcast by the base station in the downlink frame. Transmission permission probabilities are calculated on the basis of the estimated number of backlogged terminals (e.g., those needing to transmit but having no reservation).

If N_t is the number of backlogged terminals at the beginning of time slot t, in *Bayesian broadcast*, Bayes' rule is used to update, after each time slot, the estimated probability that $N_t = n$, for each nonnegative n. In pseudo *Bayesian broadcast*, it is assumed that these probability values are approximated reasonably well by a Poisson distribution. Therefore, only the mean value of the Poisson distribution is estimated. The optimum permission probability, p, in terms of minimum delay or minimum backlog, is simply the inverse of that mean value. To discriminate the QoS of different service classes, different permission probabilities can be calculated for different access classes based on the basic value p and the priority assigned to the class. Moreover, a load-based access control is included in the computation of those probabilities in order to control the multiple access interference level due to the CDMA component. In fact, the presence of MAI reduces the accuracy in the backlog estimation and can increase the dropping probability.

By controlling the contention access (transmission probabilities) and allocation (number of frames) for data services, the protocol is able to track delay requirements and dropping probabilities for different services. One of the main drawbacks of this protocol is that it is not able to support high bit rate data services or real-time services that requires the allocation in a frame of multiple slots to a single user. Explicit reservation may enhance the protocol capabilities.

2.6.4 CDMA-Oriented MAC

Most of the protocols presented in this section have been conceived in a TDMA-like scenario where the time is usually slotted. CDMA technology presents new challenges and possibilities to multimedia MAC. MAC protocols are generally part of a complex resource management mechanism performed at the base station and possibly in the network, which must not only share resources efficiently but also guarantee stability of the system, by controlling, for instance, the interference level. The peculiarity of CDMA protocols can be deeply understood by comparing the access of a new user in TDMA/FDMA systems with that in CDMA systems. In TDMA/FDMA systems, the access of a new user is allowed according to the availability of radio resources (frequency band or time slots). In a CDMA system, where the capacity is not hard-limited (*soft capacity*), the admission of a user that has to transmit at a high power level could severely degrade the performance of all other users. Therefore, a proper resource management mechanism should include specific *admission control policies* that limit the access of some users even in the presence of available radio resources. Those policies may be based on several criteria, such as the estimation of the SIR [94] or the interference-load. Alternatively, a reduction of the transmission rate of either the new user or other users already admitted in the system could make possible the admission of the new user. This can be achieved by upper layer functionalities. It is evident that a broader vision of the whole network is needed.

As an example of MAC protocol designed for CDMA-based interfaces, the MAC protocol proposed for supporting real-time and non-real-time services in the WCDMA standard [95] is considered.

2.6.4.1   WCDMA MAC

The WCDMA MAC protocol is specially designed for the WCDMA physical layer adopted by European and Japanese standardization bodies for third generation systems. In what follows, the scheme proposed for the packet data services and real-time services are described. Finally, an example on how to support mixed services is shown.

Packet data services

The state diagram of WCDMA MAC for packet data is shown in Figure 2.22. Two methods are proposed for data transmission. The first method envisages that packets are transmitted on a random access channel (RACH) in an ALOHA fashion. The mode is efficient for transmitting short and infrequent packets with no delay and minimum overhead. The mobile station does not use fast power control on this channel. As a consequence, an unacceptable interference level in highly loaded conditions and an extensive use of the RACH channels can take place. Those issues certainly deserve further evaluation. The second transmission method envisages that the mobile station requests a dedicated channel (DCH). The mobile station is thus allocated a dedicated code and transmissions are fast power controlled. The request is usually sent on the RACH. When the mobile station is already using a dedicated channel for another service, the request can be sent on the DCH. The second mode is more suitable for long packets with respect to the overhead caused by the reservation mechanism. The request message contains information about the type of service required, along with the length of the packet to be transmitted. The network responds with a resource allocation message transmitted through the common downlink forward access channel (FCH). The message indicates a set of possible transmission formats (TFs). A transmission format specifies transmission rate, code, interleaving length, and repetition/puncturing scheme (for rate matching). If the load is low, the system also indicates the specific TF and the

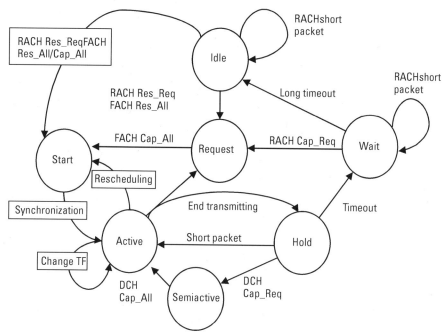

Idle: No available DCH
Start: The MS has packets to transmit; it has been allocated a time for transmission and a TF out of a TF set
Request: The MS has packets to transmit: it has been allocated a TF set
Wait: The MS has been allocated a TF set
Active: The MS is transmitting
Semiactive: No transmission; maintaining the link; waiting for a scheduling message from the network
Hold: No transmission; maintaining the link

RACHshort packet: Transmission of packets on the RACH
RACH Res_Req: Resource Request message on RACH
RACH Cap_Req: Capacity Request message on RACH
FACH Res_All: Resource Allocation message on FACH
FACH Cap_All: Capacity Allocation message on FACH
DCH Cap_All: Capacity Allocation message on DCH
DCH Cap_Req: Capacity Request message on DCH

Figure 2.22 The state diagram of the WCDMA MAC for packet data.

scheduled time for the transmission. Otherwise, before using the dedicated channel, the user must issue another request in order to get this information. After receiving the information, the user can start the transmission in the scheduled time. It is possible that the transmission is rescheduled. This can occur when a transmission of another user is prioritized. Rescheduling can be done by changing either the TF or the transmission time. Once a user is transmitting, transmission rate can be increased or decreased depending on the network load. The network transmits a rate change message, which is a pointer to another TF out of the allocated TF set. After the transmission is finished, the link is still maintained (e.g., power control commands are still sent) for a time interval, called the *holding time*. If a new packet is generated during this time, the user can start transmitting immediately with the TF used in the last transmission. However, if the generated packet is too large, the user will first transmit a message to indicate the packet length. If no packets are generated for this time, the link will be lost, but the terminal will keep the TFs for a certain period in order to facilitate future transmissions. If no packet has been generated for a long time, the mobile station will go to the idle state again, in which the user may initiate a transmission just by transmitting on the RACH.

Real-Time Services

For real-time services the allocation procedures are very similar to the data transmissions case. The first difference is that the user can start transmitting as soon as it gets the set of TFs from the network, using any TF out of the this set. In this way the user terminal can support variable bit rate such as speech. On the other hand, the network can limit the set of TFs a user can use for its transmission with respect to the initial set. This can be done when the network load is high, in order to limit the capacity of the user.

Mixed Services

The physical layer is capable of multiplexing bit streams originating from different services. Different QoS requirements can be met at physical layer by using different type of coding. However, this process has to be controlled by the MAC layer. For example, if the mobile station wishes to transmit data of a real-time service such as speech and a packet data service, it is assigned a set TFs per each of the services. The terminal is able to use any TF for real-time services, while the network assigns a specific TF for data services, according to some controlling criteria, for instance, by putting thresholds on the maximum power/rate. Such a control must be exercised because if the mobile station chooses TFs corresponding to the maximum bit rate in both the real-time and the data service, the aggregate output might be too high.

The stability of the protocol depends on higher layer-functionalities such as call admission control (CAC) and congestion control (CC). For example, when the cell is highly loaded and current users suffer a link degradation due to excessive interference, new calls are blocked (CAC) and CC orders MAC to decrease the transmission rates (lower profile TFs) of the terminals already connected to the base station. Unfortunately, these higher layer functionalities are still not completely defined.

2.6.5   Trends in MAPs Design

The previously presented protocols do not properly address the important issue of energy consumption and ignore the behavior of the wireless channel mainly by assuming that it is completely "hidden" by the physical layer interface.

The main power consumption sources from the access protocol perspective include the radio transceiver and the CPU. Maximum power is consumed in transmit mode, followed by the receive mode and standby mode. One of the options for an energy-efficient MAC design is the minimization of the number of transitions. Furthermore, due to the high power consumption for data transmission, there is a desire to minimize unsuccessful actions of the transceiver such as collisions and errors that result in retransmissions. From this point of view, MAC protocols that use reservation and polling are more attractive even if in some of them the receiver has to be turned on for longer periods. Furthermore, valuable channel resources can be allocated inefficiently during the time the channel is impaired. If transmissions are avoided during periods of bad error conditions by scheduling no traffic during these periods, energy saving can be achieved [86, 96–98].

Key elements of a proper MAC layer design as the exploitation of the peculiarities and knowledge of the wireless channel are emerging, both for improving the channel utilization and reducing the energy consumption [86, 99]. The exploitation of the capture effect [22, 23] in some receivers is an example of the fact that the fading can be properly managed to improve the performance (see Section 2.3.5). Fading can selectively disable a subset of the contending users, thus allowing a group of users to efficiently access a single base station. Furthermore, the correlation in channel errors could be exploited for reducing the energy consumption by a proper design of error recovery mechanisms at link layer [e.g., automatic repeat request (ARQ)] [100].

The knowledge of the channel status can increase the capacity in CDMA systems by transmitting mostly when the channel is good. In particular, channel-sensitive scheduling policies in MAC protocols can dramatically improve the channel utilization efficiency of MAC protocols. In [101], it is shown that, assuming a perfect knowledge of the channel information, for a 10% packet error rate and with 0.01 as the packet dropping objective, channel-aware scheduling allows admitting to the system at least twice as many users. Following this approach, power-efficient MAC protocols with channel-dependent resource allocation mechanism have been proposed in [102, 103]. Power control commands represent an implicit link quality feedback from the receiver that may be used by the MAC layer protocols to drive the scheduling process [104].

If a proper design of MAC protocols requires the study of the interaction with physical layer processes, it is also true that a proper design of lower layers should take into account MAC layer mechanisms. In [105], the particular interaction of the scheduling process with the transmit diversity process is studied. Transmit diversity has been introduced at the physical layer to improve the overall capacity of the system by improving the downlink per-user performance. Results show that the combined performance of scheduling and transmit diversity does not lead to the same conclusions on the capacity improvements as a physical layer–only study of the transmit diversity. The study of these interactions among different layers of the protocol stack requires a proper modeling of the error process. For instance, a characterization of the radio channel based only on the BER is incomplete to predict the performance of MAC protocol. Higher-order statistical information is needed [106–108]. The study of the effect of wireless channel errors on the performance of protocols represents an interesting and still open research area.

References

[1]   Rom, R., and M. Sidi, *Multiple Access Protocols Performance and Analysis*, New York: Springer-Verlag, 1990.

[2]   Prasad, R., *CDMA for Wireless Personal Communications*, Norwood, MA: Artech House, 1996.

[3]   Prasad, R., *Universal Wireless Personal Communications*, Norwood, MA: Artech House, 1998.

[4]   Spragins, J. D., J. L. Hammond, and K. Pawlikowski, *Telecommunications Protocols and Design*, Reading, MA: Addison-Wesley, 1991.

[5]   Schwartz, *Computer-Communication Network Design and Analysis*, Englewood Cliffs, NJ: Prentice-Hall, 1977.

[6]   Walrand, J., *Communication Networks,* New York: McGraw Hill, 1998.

[7]   Lee, W. C. Y., *Mobile Communications Design Fundamentals*, New York: Wiley Interscience, 1993.

[8]   Stuber, G. L., *Principles of Mobile Communication*, Boston: Kluwer Academic Publisher, 1996.

[9]   Schiller, J., *Mobile Communications*, Reading, MA: Addison-Wesley, 2000.

[10] Abramson, N., "The ALOHA System—Another Alternative for Computer Communications," *Proc. Fall Joint Computer Conference (AFIPS) 37*, 1970 pp. 281–285.

[11] Abramson, N., "Multiple Access in Wireless Digital Networks," *IEEE Proceedings*, Vol. 82, Sept. 1994, pp. 1360–1370.

[12] Kleinrock, L., and F. A. Tobagi, "Packet Switching in Radio Channels, Part I—Carrier Sense Multiple Access Nodes and Their Throughput-Delay Characteristics," *IEEE Transactions on Communications*, Vol. 23, Dec. 1975, pp. 1400–1416.

[13] Kleinrock, L., and F. A. Tobagi, "Packet Switching in Radio Channels, part II—The Hidden Terminal Problem in Carrier Sense Multiple-Access and the Busy Tone Solution," *IEEE Transactions on Communications*, Vol. 23, Dec. 1975, pp. 1417–1433.

[14] Monks, J. P., et al., "Transmission Power Control for Multiple Access Wireless Packets Networks," *IEEE Proc. Local Computer Network*, LCN 2000, pp. 12–21.

[15] Krebs, J., and T. Freeburg, "Method and Apparatus for Communicating Variable Length Messages Between a Primary Station and Remote Stations at a Data Communication System," U.S. Patent No. 4519068, 1985.

[16] Prasad, R., "Throughput Analysis of Non-Persistent Inhibit Sense Multiple Access in Multipath Fading and Shadowing Channels," *European Transactions on Telecommunications and Related Technologies*, Vol. 2, May-June 1991, pp. 313–317.

[17] Prasad, R., "Performance Analysis of Mobile Packet Radio Networks in Real Channels with Inhibit Sense Multiple Access," *IEE Proceedings*, Vol. 138, No. 5, Oct. 1991, pp. 458–464.

[18] Prasad, R., and C. Y. Liu, "Throughput Analysis of Some Mobile Packet Radio Protocols in Rician Fading Channels," *IEE Proceedings*, Vol. 138, No. 5, Oct. 1991, pp. 458–464.

[19] Hajek, B., A. Krishna, and R. O. LaMaire, "On the Capture Probability for Large Number of Stations," *IEEE Transactions on Communications*, Vol. 45, No. 2, Feb. 1997.

[20] LaMarie, R. O., A. Krishna, and M. Zorzi, "On the Use of Transmitter Power Variations To Increase Throughput in Multiple Access Radio Systems," *Wireless Networks*, Vol. 4, No. 3, June 1998, pp. 263–277.

[21] Widipangestu, I., A. J. 't jong, and R. Prasad, "Capture Probability and Throughput Analysis of Slotted ALOHA and Unslotted np-ISMA in a Rician/Rayleigh Environment," *IEEE Transactions on Vehicular Technology*, Vol. 43, Aug. 1994, pp. 457–465.

[22] Borgonovo, F., et al., "Capture Division Packet Access—A New Cellular Architecture for Future PCNs," *IEEE Communications Magazine*, Vol. 34, No. 9, Sept. 1996, pp. 154–162.

[23] Davis, D. H., and S. A. Gronemeyer, "Performance of Slotted ALOHA Random Access with Delay Capture and Randomized Time Arrival," *IEEE Transactions on Communications*, Vol. 28, May 1980, pp. 703–710.

[24] Zorzi, M., and R. R. Rao, "Capture and Retrasmission Control in Mobile Radio," *IEEE Journal on Selected Areas on Communications*, Vol. 12, Oct. 1994, pp. 1289–1298.

[25] Hajek, B., A. Krishna, and R. O. LaMaire, "On the Capture Probability for a Large Number of Stations," *IEEE Transaction on Communications*, Vol. 45, Feb. 1997, pp. 254–260.

[26] Katzela, I., and M. Naghineh, "Channel Assignment Schemes for Cellular Mobile Telecommunication Systems: A Comprehensive Survey," *IEEE Personal Communications*, Vol. 3, No. 6, 1996, pp. 39–47.

[27] Hua Jiang, and S. S. Rappaport, "A Channel Borrowing Scheme for TDMA Cellular Communication Systems," *Proc. IEEE Vehicular Technology Conference*, 1995, pp. 97 –101.

[28] Maric, S. V., E. Alonso, and G. Metivier, "Adaptive Borrowing of Ordered Resources for the Pan-European Mobile Communication (GSM) System," *IEE Proceedings*, Vol. 141, No. 2, April 1994, pp. 93 –97.

[29] Koo, B., and V. K., Jain, "Efficient Channel Borrowing Strategy in PCN Systems," *Proc. 2nd International Conference on Universal Personal Communications*, 1993, pp. 570–574.

[30] Zander, J., "Performance of Optimum Transmitter Power Control in Cellular Radio Systems," *IEEE Transactions on Vehicular Technology*, Vol. 41, No. 1, Feb. 1992, pp. 57–66.

[31]   Roberts, L. G., "Dynamic Allocation of Satellite Capacity Through Packet Reservation," *Proc. National Computer Conference (AFIPS)*, June 1973, pp. 711–716.

[32]   Crowther, W., et al., "A System for Broadcast Communications: Reservation-ALOHA," in *Proc. 6th Hawaii Int. Conf. Syst. Sci*, 1973, pp. 596–603.

[33]   Schwartz, *Computer-Communication Network Design and Analysis*, Englewood Cliffs, NJ: Prentice-Hall, 1977.

[34]   Goodman, D. J., and S. X. Wei, "Efficiency of Packet Reservation Multiple Access," *IEEE Transaction Vehicular Technology*, Vol. 40, Feb. 1991, pp. 170–176.

[35]   Cheung, J. C. S., et al., "Effect of PRMA on Speech Objective Quality," *IEEE Electronic Letters*, Vol. 29, No. 2, Jan. 1993, pp. 152–153.

[36]   Hanzo, L., et al., "A Packet Reservation Multiple Access Assisted Cordless Telecommunications Scheme," *IEEE Transactions Vehicular Technology*, Vol. 23, No. 2, May 1994, pp. 234–244.

[37]   Verdu, S., *Multiuser Detection*, Cambridge, UK: Cambridge University Press, 1998.

[38]   Proakis, J. G., *Digital Communications*, Fourth Edition, New York: McGraw Hill, 2001.

[39]   Dinan, H. E., and B. Jabbari, "Spreading Codes for Direct Sequence CDMA and Wideband CDMA Cellular Networks," *IEEE Commun. Mag.*, Vol. 36, Sept. 1998, pp. 48–54.

[40]   De Gaudenzi, R., C. Elia, and R. Viola, "Bandlimited Quasi-Synchronous CDMA: A Novel Satellite Access Technique for Mobile and Personal Communication System," *IEEE J. Select. Areas Commun.*, Vol. 10, No. 2, Feb. 1992, pp. 328–343.

[41]   Lehnert, J. S., and M. B. Pursley, "Error Probabilities for Binary Direct Sequence Spread Spectrum Communications with Random Signature Sequences," *IEEE Transactions Communications*, Vol. COM-35, Jan. 1987, pp. 85–96.

[42]   Verdu, S., "Multi-User Detection with Random Spreading and Error Correction Codes: Fundamental Limits," *Proc. Allerton Conference*, 1997.

[43]   Verdu S., and S. Shamai, "Spectral Efficiency of CDMA with Random Spreading," *IEEE Transactions on Information Theory*, Vol. 45, No. 2, 1999, pp. 622–640.

[44]   Jakes, *Microwave Mobile Communications*, New York: John Wiley & Sons, 1974.

[45]   Poor, H. V., and G. W. Wornell, *Wireless Communications-Signal Processing Perspective*, Englewood Cliffs, NJ: Prentice Hall, 1998.

[46]   Verdu, S., "Optimum Multiuser Asymptotic Efficiency," *IEEE Transaction on Communications*, Sept. 1996, pp. 890–897.

[47]   Zihua, G., and K. B. Letaief, "An Effective Multiuser Receiver for DS/CDMA Systems," *IEEE Journal on Selected Areas in Communications*, Vol. 19, No. 6, June 2001, pp. 1019–1028.

[48]   Shama, Y., B. R. Vojcic, and B. Vucetic, "Suboptimum Soft-Output Detection Algorithms for Coded Multiuser Systems," *IEEE Transactions on Communications*, Vol. 48, No. 10, Oct. 2000, pp. 1622–1625.

[49] Tang, Z., Z. Yang, and Y. Yao, "Robust Blind Adaptive Multiuser Detector," *Electronics Letters*, Vol. 35, No. 5, March 4, 1999, pp. 384–385.

[50] Tulino, A. M., and S. Verdu, "Asymptotic Analysis of Improved Linear Receivers for BPSK-CDMA Subject to Fading," *IEEE Journal on Selected Areas in Communications*, Vol. 19, No. 8, Aug. 2001, pp. 1544–1555.

[51] Chen, W., U. Mitra, "An Improved Blind Adaptive MMSE Receiver for Fast Fading DS-CDMA Channels," *IEEE Journal on Selected Areas in Communications*, Vol. 19, No. 8, Aug. 2001, pp. 1531–1543.

[52] Hanly, S. V., and D. Tse, "Resource Pooling and Effective Bandwidths in CDMA Networks with Multiuser Receivers and Spatial Diversity," *IEEE Transactions on Information Theory*, Vol. 47, No. 4, May 2001, pp. 1328–1351.

[53] Honig, M., and M. K. Tsatsanis, "Adaptive Techniques for Multiuser CDMA Receivers," *IEEE Signal Processing Magazine*, Vol. 17, No. 3, May 2000, pp. 49–61.

[54] Zhang, J., E. K. P. Chong, "Power Control for Spread-Spectrum Networks in Fading Channels," Proc. *IEEE Sixth International Symposium on Spread Spectrum Techniques and Applications*, ISSSTA'2000, Parsippany, New Jersey, Sept. 2000.

[55] Viterbi, A. M., and J. Viterbi, "Erlang Capacity of a Power Controlled CDMA System," *IEEE Journal on Selected Areas of Communications*, Vol. 11, Aug. 1993, pp. 892–900.

[56] Viterbi, A. M., J. Viterbi, and E. Zehavi, "Performance of Power Controlled Wideband Terrestrial Digital Communication," *IEEE Transactions on Communications*, Vol. 41, No. 4, April 1993, pp. 559–569.

[57] Cianca, E., et al., "Approach to Maximize the Capacity of a Multimedia CDMA Wireless System," *IEEE Proc. VTC'98*, May 1998, pp. 909–913.

[58] Gilhoesen, K. S., et al., "On the Capacity of a Cellular CDMA System," *IEEE Transaction on Vehicular Technology*, Vol. 40, No. 2, May 1991, pp. 303–312.

[59] Vembu, S., and A. J. Viterbi, "Two Philosophies in CDMA-A Comparison," *Proc. IEEE 46th Vehicular Technology Conference*, 1996, pp. 869–873.

[60] Prasad, R., W. Mohr, and W. Konhauser, *Third Generation Mobile Communication System*, Norwood, MA: Artech House, 2000.

[61] Calin, D., and M. Areny, "Impact of Radio Resource Allocation Policies on the TD-CDMA System Performance: Evaluation of Major Critical Parameters," *IEEE Journal on Selected Areas in Communications*, Vol. 19, No. 10, Oct. 2001, pp. 1847–1859.

[62] Park, S., K. Minjung Kim, and Nak-Myeong Kim, "An Enhanced Adaptive Time Slot Assignment Using Access Statistics in TD/CDMA TDD System," *Proc. IEEE Vehicular Technology Conference 2001*, VTC 2001 Spring, pp. 511–516.

[63] Klein, A., G. W. Kaleh, and P. W. Baier, "Zero Forcing and Minimum Mean-Square Error Equalization for Multiuser Detection in Code Division Multiple Access Channels," *IEEE Transaction on Vehicular Technology*, Vol. 45, No. 2, May 1996, pp. 276–287.

[64] Klein, A., and P. W. Baier, "Linear Unbiased Data Estimation in Mobile Radio Systems Applying CDMA," *IEEE Journal on Selected Areas in Communications*, Vol. 11, Sept. 1993, pp. 1058–1066.

[65]   Kaleh, G. K., "Channel Equalization for Block Transmission Systems," *IEEE Journal on Selected Areas in Communications*, Vol. 13, Jan. 1995, pp. 110–120.

[66]   Jung, P., and J. Blanz, "Joint Detection with Coherent Receiver Antenna Diversity in CDMA Mobile Radio Systems," *IEEE Transaction on Vehicular Technology*, Vol. 44, Feb. 1995, pp. 76–88.

[67]   Klein, A., B. Steiner, and A. Steil, "Known and Novel Diversity Approaches as a Powerful Means to Enhance the Performance of Cellular Mobile Radio Systems," *IEEE Journal on Selected Areas in Communications*, Vol. 14, Dec. 1996, pp. 1784–1795.

[68]   Benvenuto, N., and G. Sostrato, G., "Joint Detection with Low Computational Complexity for Hybrid TD-CDMA Systems," *IEEE Journal on Selected Areas in Communications*, Vol. 19, No. 2, Feb. 2001, pp. 245–253.

[69]   Vollmer, M., M. Haardt, and J. Gotze, "Comparative Study of Joint-Detection Techniques for TD-CDMA Based Mobile Radio Systems," *IEEE Journal on Selected Areas in Communications*, Vol. 19, No. 8, Aug. 2001, pp. 1461–1475.

[70]   Prasad, R., and S. Hara, "An Overview of Multicarrier-CDMA," *IEEE Communications Magazine*, Vol. 35, No. 12, Dec. 1997, pp. 126–133.

[71]   Casal, C. R., F. Schoute, and R. Prasad, "A Novel Concept for Fourth Generation Mobile Multimedia Communication," *Proc. IEEE Vehicular Technology Conference*, 1999. VTC 1999–Fall. IEEE VTS 50th , Vol. 1, 1999, pp. 381–385.

[72]   Morinaga, N., M. Nakagawa, and R. Kohno, "New Concept and Technologies for Achieving Highly Reliable and High Capacity Multimedia Wireless Communications Systems," *IEEE Communications Magazine*, Jan. 1997, pp. 34–40.

[73]   Bingham, J. A. C., "Multicarrier Modulation for Data Transmission: An Idea Whose Time Has Come," *IEEE Communications Magazine*, Vol. 28, No. 5, May 1990, pp. 5–14.

[74]   Fazel, K., and L. Papke, "On the Performance of Convolutionally Coded CDMA/OFDM for Mobile Communication System," *Proc. IEEE PIMRC'93*, Yokohama, Japan, Sept. 1993, pp. 468–472.

[75]   Hara, S., T.-H. Lee, and R. Prasad, "BER Comparison of DS-CDMA and MC-CDMA for Frequency Selective Fading Channels," *Proc. of 7th Tyrrhenian International Workshop on Digital Communications*, Viareggio, Italy, Sept. 1995, pp. 3–14.

[76]   Sourour, E. A., and N. Nakagawa, "Performance of Orthogonal Multicarrier CDMA in a Multipath Fading Channel," *IEEE Transactions Communications*, Vol. 44, March 1996, pp. 356–366.

[77]   Yee, N., J. P. Linnartz, and G. Fettweis, "Multi-Carrier CDMA in Indoor Wireless Radio Network," *IEICE Transaction Communications*, Vol. E77-B, July 1994, pp. 900–904.

[78]   Fazel, K., S. Kaiser, and M. Schnell, "Flexible and High Performance Cellular Mobile Communications System Based on Orthogonal Multicarrier SSMA," *Wireless Personal Communications*, Vol. 2, Nov. 1995, pp. 121–144.

[79]   Van Nee, R., and R. Prasad, *OFDM for Wireless Multimedia Communications*, Norwood, MA: Artech House, 2000.

[80] Weinstein, S. B., and P. M. Ebert, "Data Transmission by Frequency Division Mutliplexing Using the Discrete Fourier Transform," *IEEE Trans. Comm.*, Vol. COM-19, Oct. 1971, pp. 628–634.

[81] Zou, W. Y., and Y. Wu, "COFDM: An Overview," *IEEE Trans. Broadc.*, Vol. 41, No. 1, March 1995, pp. 1–8.

[82] Cimini, L. J., Jr., "Analysis and Simulation of a Digital Mobile Channel Using Orthogonal Frequency Division Multiplexing," *IEEE Transaction on Communications*, Vol. COM-33, No. 7, July 1985, pp. 665–675.

[83] Mikkonen, J., et al., "The Magic WAND-Functional Overview," *IEEE JSAC*, Vol. 12, No. 6, Feb. 1997, pp. 28–32.

[84] Chuang, J., and N. Sollenberger, "Beyond 3G: Wideband Wireless Data Access Based on OFDM and Dynamic Packet Assignment," *IEEE Communication Magazine*, July 2000, pp. 78–87.

[85] Kondo, S., and L. B. Milstein, "Performance of Multi-Carrier DS-CDMA Systems," *IEEE Transactions Communications*, Vol. 44, No. 2, Feb. 1996, pp. 356–366.

[86] Havinga, P. J. M., and G. J. M. Smit, "Energy-Efficient Wireless Networking for Multimedia Applications," *Wireless Communications and Mobile Computing*, Vol. 1, No. 2, April/June 2001, pp. 165–184.

[87] Akyildiz, I. F., et al., "Medium Access Control Protocols for Multimedia Traffic in Wireless Networks," *IEEE Network*, July/Aug. 1999, pp. 39–47.

[88] Sanchez, J., R. Martinez, and M. Marcellin, "A Survey of MAC Protocols Proposed for Wireless ATM," *IEEE Network*, Nov./Dec. 1997, pp. 52–62.

[89] Tan, L., and Q. Zhang, "A Reservation Random-Access Protocol for Voice/Data Integrated Spread Spectrum Multiple-Access Systems," *IEEE Journal on Selected Areas of Communications*, Vol. 14, Dec. 1996, pp. 1717–1727.

[90] Dyson, D. A., and Z. J. Haas, "The Dynamic Packet Reservation Multiple Access Scheme for Multimedia Traffic," ACM/Baltzer, *Journal Mobile Networks and Applications*, 1999.

[91] Bianchi, G., F. Borgonovo, and L. Fratta, "CPRMA: A Centralized Packet Multiple Access for Local Wireless Communications," *IEEE Transactions on Vehicular Technology*, Vol. 46, No. 2, May 1997, pp. 422–436.

[92] Brand, A. E., and H. Aghvami, "Multidimensional PRMA (MD-PRMA): A Versatile Medium Access Strategy for the UMTS Mobile to Base Station Channel," in *Proc. PIMRC'97*, Helsinki, Finland, Sept. 1997, pp. 524–528.

[93] Brand, A. E., and H. Aghvami, "Multidimensional PRMA with Prioritized Bayesian Broadcast—A MAC Strategy for Multiservice Traffic over UMTS," *IEEE Transactions Vehicular Technology*, Vol. 47, No. 4, Nov. 1998.

[94] Zhao, L., and M. El Zarki, "SIR-Based Call Admission Control for DS-CDMS Cellular Systems," *IEEE Journal on Selected Areas on Communications*, Vol. 12, No. 4, May 1994, pp. 638–644.

[95]   Roobol, C., et al., "A Proposal for an RLC/MAC Protocol for Wideband CDMA Capable of Handling Real Time and Non Real Time Services," in *Proc. IEEE VTC'98*, Ottawa, Canada, May 1998, pp. 107–111

[96]   Cianca, E., et al., "Power Management in IP-based CDMA Mobile Systems," *in Proc. WPMC00*, Bangkok, Nov. 2000.

[97]   Cianca, E., M. Ruggieri, and R. Prasad, "Improvement of TCP/IP Performance over CDMA Wireless Links: A Physical Layer Approach," in *Proc. IEEE PIRMC01*, San Diego, California, Sept. 2001, pp. 83–87.

[98]   Akyildiz, I. F., D. A. Levine, and I. Joe, "A Slotted CDMA Protocol with BER Scheduling for Wireless Multimedia Networks," *IEEE/ACM Transactions on Networking*, Vol. 7, No. 2, April 1999, pp. 146–159.

[99]   Zorzi, M., and R. R. Rao, "Perspectives on the Impact of Error Statistics on Protocols for Wireless Networks," *IEEE Personal Communications*, Oct. 1999, pp. 32–40.

[100] Zorzi, M., R. R. Rao, and L. B. Milstein, "Error Statistics in Data Transmission over Fading Channels," *IEEE Transaction on Communications,* Vol. 46, Nov. 1998, pp. 1468–1477.

[101] Zorzi, M., and R. R. Rao, "The Role of Error Correlations in the Design of Protocols for Packet Switched Services," *Proc. 35th Annual Allerton Conf.*, Sept. 1997, pp. 749–758.

[102] Shauh, J., "Resource Control for Wireless Data Services," Ph.D. Dissertation, UCSD, 1998.

[103] Bhagwat, P., et al., "Using Channel State Dependent Packet Scheduling to Improve TCP Throughput over Wireless LANs," *Wireless Networks*, 1997, pp. 91–102.

[104] Mandyam, G. D., and Y.-C. Tseng, "Packet Scheduling in CDMA Systems Based on Power Control Feedback," in *Proc. IEEE International Conference on Communications*, 2001, ICC 2001, Vol. 9, 2001, pp. 2877–2881.

[105] Kogiantis, A. G., N. Joshi, and O. Sunay, "On Transmit Diversity and Scheduling in Wireless Packet Data," *Proc. IEEE ICC'01*, 2001, pp. 2433–2437.

[106] Zorzi, M., R. R. Rao, L. B. Milstein, "On the Accuracy of a First-Order Markov Model for Data Block Transmission on Fading Channels," in *Proc. IEEE ICUPC'95*, Nov. 1995, pp. 211–215.

[107] Zorzi, M., and R. R. Rao, "Lateness Probability of a Retransmission Scheme for Error Control on a Two-State Markov Channel," *IEEE Transaction on Communications*, Oct. 1999, pp. 1537–1548.

[108] Zorzi, M., and R. R. Rao, "On the Statistics of Block Errors in Bursty Channels," *IEEE Transaction on Communications*, June 1997, pp. 660–667.

3

IP Network Issues

3.1 Introduction

The evolution towards a common, flexible, and seamless IP-based core network that will connect heterogeneous networks gives rise to several issues at the network layer. The need for ensured QoS is the key of this evolution. Fundamentally, the day that packet-switched networks can credibly approach the QoS of circuit-switched networks is the day customers stop paying for two networks. The QoS problem involves integrating delay-sensitive applications such as voice audio and video onto a single network with delay-insensitive applications, such as e-mail, fax, and static file transfer. That network must be able to discriminate, differentiate, and deliver communications, content, and commerce services. Moreover, it needs to support communication services creation, modification, bundling, and billing in a way that is unobtrusive yet powerful for end users, especially business users. The scenario becomes more challenging as QoS and security issues have to be faced in a mobile environment. The Internet has not been designed with mobility in mind and lacks mechanisms to support mobile users. Some architectures have been proposed for supporting mobility in the Internet. The most important is Mobile IP, which is discussed in Section 3.3, mainly referring to Mobile IP version 4. Guidelines behind mobility management are discussed in Section 3.2. In the rest of the chapter, QoS provision, security, and routing issues have been addressed. Each section first discusses the main approaches generally proposed for IP fixed networks and then introduces some extensions for mobile networks. Specifically, Section 3.4 discusses packet route optimization through proper routing algorithms. Section 3.5 introduces the two basic architectures developed to support QoS in IP networks, such as Integrated Service (IntServ) and Differentiated Service (DiffServ). Moreover, the Multi-Protocol Label Switching (MPLS) is presented as a promising technology for delivering QoS on IP-based networks. Protocols and mechanisms to fulfill security needs of new multimedia users in IP networks are addressed in Section 3.6. In particular, IPSec protocol, AAA architecture, and the extension for Mobile IP are presented. Finally, proposals to solve some of the open issues in Mobile IP are described in Section 3.7. Conclusions are drawn in Section 3.8.

3.2   Mobility Management

When a mobile node is roaming through one or more service areas, mobility management mechanisms are required to locate it for call delivery and maintenance of its connections. Generally in cellular systems, mobility management is performed through two main mechanisms:

1. *Location management*, which is used for discovering the current attachment point of the mobile user for call delivery. It consists of two phases. In the first phase, called the location *registration* (or *location update*), the mobile terminal periodically notifies the network for its new access point, allowing the network to authenticate the user and revise the user's location profile. The second phase is *call delivery*. The network is queried for the user location profile and the current position of the mobile host is found.

2. *Handoff management*, which enables the user to keep its connection alive as it moves and changes its access point to the network. This is performed in three steps: *initiation, connection generation,* and *data flow control.*

When the user moves within a service area or cell and changes communications channels allocated by the same BS, the mobility management procedure is called *intracell* handoff. *Intercell* handoff occurs when the user moves into an adjacent service area or cell for which all mobile connections are transferred to a new BS. While performing handoff, a mobile terminal can be simultaneously connected to two or more BSs and use some kind of signaling diversity to combine the multiple signals. This condition is called *soft handoff.* In *hard* handoff, the mobile device switches from one base station to another with active data being forwarded on only one path at a time.

3.2.1   Mobility Classes

In general, there are four categories that support IP mobility:

1. *Picomobility* is the movement of a mobile node (MN) within the same BS. The operating space is the space around person that typically

extends up to 10m in all directions and envelops the person being either stationary or in motion.

2. *Micromobility* is the movement of an MN within or across different BSs within a subnet and occurs very rapidly. Management of micromobility is accomplished using link-layer support (layer 2 protocol) already implemented in existing cellular networks.

3. *Macromobility* is the movement of an MN across different subnets within a single domain or region, and occurs relatively less frequently. This is currently handled by Internet mobility protocols such as Mobile IP.

4. *Global mobility* is the movement of an MN among different administrative domains or geographical regions. This is also handled by layer 3 techniques such as Mobile IP.

Mobility management has a responsibility of providing uninterruptible connectivity during micro- and macromobility, which usually occur over relatively short timescales. Global mobility, on the other hand, usually involves longer timescales. Therefore, the goal is just to ensure that mobile users can reestablish communication after they change the domain, but not necessarily to provide uninterruptible connectivity.

3.2.2 Architectures for Mobility Supporting

There are several frameworks that support mobile users, and the Internet Engineering Task Force (IETF) standardizes two of them: Mobile IP and Session Initiation Protocol (SIP).

- Mobile IP [1] supports application-layer transparent IP mobility. The basic Mobile IP protocol does not require protocol upgrades in stationary correspondent nodes (CNs) and routers. Its drawback is that it does not consider the integration of additional functions such as authentication and billing, which are critical for successful adoption in commercial networks.

- SIP [2] is an application-layer control (signaling) protocol that can establish, modify, and terminate multimedia sessions or calls. The main disadvantage of SIP mobility is that it cannot support TCP connections and is also not an appropriate solution for micro- or macromobility.

SIP mobility will not be a subject of further discussion in this chapter, which will focus on Mobile IP.

3.3   Mobile IP

Mobile IP refers to a set of protocols, developed and still under development by the IETF to allow the Internet Protocol to support the mobility of a node [1]. The idea for Mobile IP first emerged in 1995 and since then it has undergone some changes.

First of all, it is necessary to understand what makes the IP mobility a hard task. An IP address consists of two parts: a prefix that identifies the subnet in which the node is located and a part that identifies the node within the subnet. Routers use look-up tables to forward incoming packets according to their destination addresses. A router does not store the addresses of all computers in the Internet, which is not feasible. Only prefixes are stored in the routing tables and some optimizations are applied. Therefore, as a receiver moves outside the original subnet, it cannot be reached any longer. A new IP address should be assigned to the mobile node, but this operation requires time. Specifically, the Domain Name System (DNS) needs time to update its internal tables for mapping a logical name to an IP address. Therefore, this approach cannot work if the node moves quite often. Moreover, a browser and a Web server being, respectively, client and server in communicating, generally use the TCP/IP protocol suite to establish a reliable end-to-end communication. Using TCP, a virtual connection must be established before data transmission and reception. When a logical circuit is created, both connection sides must be assigned port numbers, in order to let application layers keep track of the communications. Every end-to-end TCP connection is identified by four values: client IP address, client TCP port, server IP address, and server TCP port. These values must be constant during the conversation. Therefore, a TCP connection will not survive any address change.

Allowing the mobile nodes the use of two IP addresses provides the solution to the above problem. In Mobile IP, *home address* remains unchanged regardless of where the node is attached to the Internet and is used to identify TCP/IP connections. Another IP address, which is dynamic, is assigned to the mobile node when it moves to another network different than its home network. This address is called the *care-of address* (COA) and is used to identify the mobile node point of attachment in the network topology. Using its home address, the mobile node is seen by other Internet hosts as a part of its own home network. In other words, other Internet hosts do not have to know the actual mobile node location. The *home agent* (HA) is located in the home network. Whenever the mobile node is not attached to its home network but to another network, called the *foreign network*, the home agent gets all packets destined for the mobile node and arranges their delivering to the *foreign agent* (FA). The FA then sends the packets to the mobile node. An HA can be implemented on a router that is responsible for the home network. This position is quite appropriate because without any optimization to mobile IP, all packets for the

MN have to go through the router anyway. An alternative solution consists of making the router behave as a manager for MNs belonging to a virtual home network. In this case, all MNs are always in a foreign network. The HA could also be implemented on an arbitrary node in the subnet. This solution is necessary when the router software cannot be changed, but it has the disadvantage that there is a double-crossing of the router by the packet if the MN is in a foreign network.

In general, the Mobile IP concept can be considered as a combination of three major functions:

1. *Discovery mechanism:* to allow mobile computers to determine their IP address as they move from network to network;
2. *Registering* of the new IP address with its home agent;
3. *Tunneling* (delivery) of packets to the new IP address of the mobile node.

In the following sections, more details about the mobile operations for each function [3] are provided, referring to the mobile IP version 4 (IPv4).

3.3.1 Mobile IPv4

Mobility agents (home and foreign agents) advertise their availability on each link where they provide service. They typically broadcast agent advertisements, at regular intervals (one or few second intervals) and in random fashion. In this way, the MN is able to know that it has moved and can find a foreign agent. In Mobile IPv4, these advertisement messages are an extension to well known Internet Control Message Protocol (ICMP) Router Discovery messages specified in RFC 1256 [4]. These messages are known as *agent advertisements.*

3.3.1.1 Discovery Mechanism

The agent advertisements perform the following functions:

- To detect home agents and foreign agents;
- To provide care-of addresses;
- To inform the mobile node about special features provided by foreign agents, e.g., alternative encapsulation techniques;
- To permit mobile nodes to determine the network number and congestion status of their link to the Internet;
- To let the mobile node know whether it is in its home network or in a foreign network by identifying whether the agent is a home agent or a foreign agent.

The agent advertisement format according to RFC 1256 and the mobility extension format are shown in Figure 3.1.

The *type* field allows mobile nodes to distinguish between the various types of extensions that may be applied by mobility agents to the ICMP router advertisements. In the ICMP part, type is set to 9. The *lifetime* field gives an information to the mobile node about the time it will be served by the HA.

The *code* field is set to 0 if the agent routes traffic from nonmobile nodes, or 16 if it does not route anything other than mobile traffic. The field *#addresses* represent the addresses advertised with this packet. Lifetime denotes the time interval for which this advertisement is valid. Preference levels for each address help a node in the choice of the router that is the most eager one to get a new node. In the mobility extension, type file is set to 16. *Length* denotes the number of COAs provided with the message. *Sequence number* shows the total number of advertisements sent since initialization. By *registration lifetime* an agent can specify the maximum lifetime in seconds a node can request during registration, as explained in Section 3.3.1.2. Flags are defined as follows:

- R: registration required. Registration with this foreign agent (or another foreign agent on this link) is required, even if using a collocated COA.

- B: The foreign agent is busy.

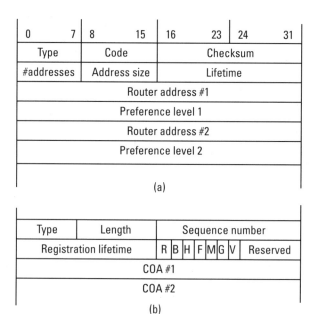

Figure 3.1 Agent advertisement format: (a) RFC 1256; and (b) mobility extension.

- H: The agent is a home agent.
- F: The agent is a foreign agent.
- M: Minimal encapsulation (RFC 2004 [5]).
- G: GRE encapsulation (RFC 1701 [6]).
- V: Van Jacobson header compression (RFC 1144[7]).

A mobile node is allowed to send ICMP router solicitation messages in order to elicit a mobility agent advertisement. If the mobile node does not receive agent advertisements any more from a foreign agent, it will assume that the foreign agent is no longer accessible.

3.3.1.2 Registration

There are two types of registration messages: registration *request* and registration *reply*, both sent to the User Datagram Protocol (UDP) port 434 [3]. UDP is used for reasons of low overhead and better performance as compared with TCP in wireless environments (see Chapter 4).

Through the registration process, a mobile node can:

- Inform the HA of the current location for correct forwarding of packets;
- Request routing services from a foreign agent on a foreign link;
- Renew a registration that is due to expire;
- Deregister when it returns to its home link.

A mobile node has to deregister with the home agent after moving to a new network.

To register its COA with the home agent, the mobile node has to send a registration request message to the home agent. The message flow diagram of the registration process is illustrated in Figure 3.2.

3.3.1.3 Registration Request

There are two possible scenarios in the registering process with respect to the COA source:

1. The mobile node gets its care-of address from a foreign agent;
2. The mobile node gets its care-of address from a DHCP server.

If the first case, the mobile node sends the request to the foreign agent, which then relays the request to the HA. In the second case, DHCP server

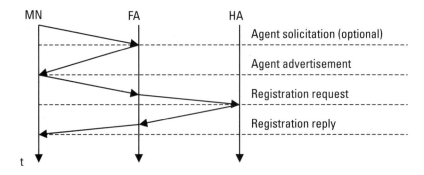

Figure 3.2 Registration signaling diagram.

assigns the mobile node a COA. Then it sends its request directly to the home agent, using its *collocated* care-of address as the source IP address of the request. The registration request has the structure illustrated in Figure 3.3.

The field type is set to 1 for a registration request. With the S bit an MN can specify if it wants the HA to retain prior mobility bindings. This allows for simultaneous bindings. The bit B is used to ask the HA to encapsulate broadcast datagrams from the home network for delivery to the COA and from there to the MN. D describes whether or not the MN is collocated with its COA. The bits M, G, and V have been already defined for the agent advertisement. The lifetime field gives information to the MN on the validity of the registration in seconds. Home address is the fixed IP address of the MN. Home address is the

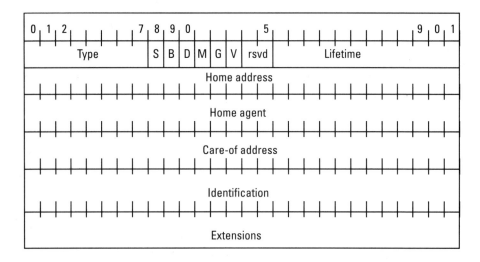

Figure 3.3 Registration request format.

IP address of the HA, and the care-of address represents the tunnel endpoint. The identification field is a 64-bit field and is used for replay protection. In fact, a malicious user could disrupt communications between the HA and MN by simply supplying a registration request containing a bogus COA: as a result, the MN would become unreachable. The method to avoid that consists of including along with the registration a value that changes every new registration. There are two ways to make this identification unique. The first way is to insert a *timestamp* into the identification field. The second way is to fill the identification field with a newly generated random number called *nonce*.

3.3.1.4 Registration Reply

The format of the registration reply is illustrated in Figure 3.4.

The code field describes the status of the registration, whether it is successful or not, and what the reasons for an eventual unsuccessful registration are. When the code number is 133, it means that a resynchronization between the home agent and the mobile node is required. This synchronization can either be time-based (timestamp), or based on the exchange of randomly generated nonce values. A more detailed description can be found in [3].

3.3.1.5 Dynamic Home Agent Discovery

If the value of the code field in the registration reply is 136, then it allows the mobile node to find the address of a home agent when required. If the registration request uses a broadcast address as a destination address, then every home agent on the home network should receive it, and reject it. But, the registration

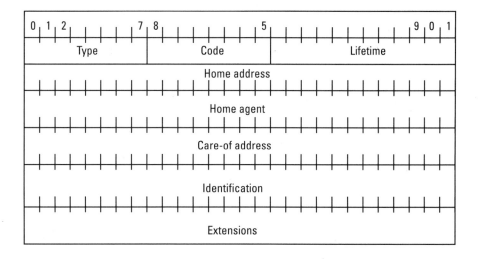

Figure 3.4 Registration reply format.

reply that contains the rejection also contains the respective home agent address, so that the mobile node can make a registration request again, and succeed.

3.3.1.6 Tunneling

After a successful registration, the home agent will begin to attract packets destined for the mobile node and tunnel them to the mobile node at its care-of address. The tunneling process is accomplished by using encapsulation techniques. The default algorithm that must always be supported is simple IP-within-IP encapsulation, as described in RFC 2003 [8].

The encapsulation of an IP packet, by preceding it with a new IP header (the tunnel header) is shown in Figure 3.5. The tunnel IP header indicates the presence of the encapsulated IP packet by using the value 4 in the outer protocol field. The inner header is not modified except decrement the time-to-live (TTL).

Other types of encapsulation techniques can also perform the tunneling procedure. These techniques are included in different encapsulation protocols such as the minimal encapsulation protocol [5] and the Generic Routing Encapsulation (GRE) protocol [6].

The minimal encapsulation protocol combines the information from the tunnel header with the information from the inner minimal encapsulation header to reconstruct the original IP header. In this way the header overhead is reduced, but the processing of the header is slightly more complicated.

In the GRE encapsulation protocol a source route entry (SRE) is provided in the tunnel header. By using the SRE, an IP source route, which includes the intermediate destinations, can be specified.

3.3.1.7 Proxy and Gratuitous ARP

When the mobile node leaves the home network, the home agent starts performing proxy Address Resolution Protocol (ARP) [9], in order to enable nodes located in the home network to communicate with the mobile. When the mobile node moves to another network (subnetwork), the home agent informs all IP nodes in the home network that the mobile node moved away. It is accomplished by sending gratuitous ARP messages. These messages update all

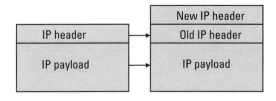

Figure 3.5 IP-within-IP encapsulation.

ARP caches of each node in the home network. After that, by using proxy ARP, the home agent intercepts the packets sent by these IP nodes to the mobile node. The intercepted packets are then tunneled to the COA.

3.4 IP Routing

The main task of routing protocols is to deliver packets from a certain source to a given destination. This must be done maintaining the correct data order and introducing the minimum delay. A good routing algorithm should be characterized by:

- Robustness, to adapt to the variations in the traffic and the availability of the links;
- Low computational complexity, to reduce the delay;
- Stability, to converge towards an acceptable solution even after changes in the topology;
- Optimality, to identify the more suitable path for each packet in terms of delay;
- Equity, to ensure that no nodes get privileges;

Paths usually are optimized with respect to the following performance metrics:

- Number of hops that the packet has to cross to reach the destination;
- Total cost of a path that is the sum of costs of each path used to reach the destination, where the cost of a path is a parameter like the delay (or the length of the queue, reliability, bandwidth).

Three basic routing techniques can be identified:

1. Routing by network address;
2. Label swapping;
3. Source routing.

In the *routing by address*, only the destination address is specified in the header of each packet. Each node has to find the next hop according to the routing table. A routing table is a set of couples (address, interface).

In the *label swapping* technique, each packet contains a label that is common to packets in the link between the same nodes. Each node replaces the current label with a new one before forwarding the packet. This technique is typical for connection-oriented protocol and in particular for delay-sensitive transmissions. In case of *source routing*, the header of the packet contains information on the whole path to the destination.

In the rest of this section, we mainly refer to *routing by address* as it is typical in connectionless protocols like IP. According to where the routing table is evaluated, a centralized and a distributed routing can be distinguished. In case of centralized routing, the evaluation of the routing table is performed in a routing control center (RCC) that stores all the information on topology, traffic conditions on each node and availability of the links. The RCC periodically updates the routing tables of each node of the network. In case of distributed routing, each node computes the routing tables using information provided by adjacent nodes.

The two main classes of distributed routing protocols are: distance vector and link state protocols [10].

3.4.1 Distance Vector Protocols

Presently the most widely used routing protocol between gateways in the Internet is the Routing Information Protocol (RIP). RIP belongs to the class of distance vector protocols, which are also referred to as Bellman-Ford algorithms. In a distance vector protocol, each node contains a routing table with a list of shortest paths to the other nodes in the network. At the moment of initializing the network, each node only has knowledge of its own address and the ability to transmit on all the links connecting it to its neighboring nodes. It broadcasts a package to all its neighboring nodes. If the distance to the node is shorter than the value already mentioned in the table, the distance table is updated with the new value. The updated distance vector is transmitted again to the neighboring nodes. When the transmission of distance vector no longer causes an update of the tables, the protocol has converged and the topology of the network has been discovered. Distance vector protocols are simple, but also offer a limited functionality.

3.4.2 Link State Protocols

In networks that use link state algorithms, all nodes maintain a distributed map of the network, the link state database. The maps are be updated quickly after a modification of the network, by means of a flooding message spread throughout the network. In contrast with the distance vector protocols, in link state protocols the nodes do not simply exchange distances. Each node is informed about the state of links towards adjacent nodes and each node transmits a link state packet to all the other nodes. The packet contains information on its adjacent nodes and the state of the related links. Nodes save the most recently received link state packets. Routing tables are evaluated by the Dijkstra algorithm, also called *shortest path first*. It converges faster than distance vector protocols. Several reasons exist to prefer a link state protocol to a distance vector protocol, despite its higher complexity [10]:

- *Support of multiple paths to a destination.* A distance vector protocol would select randomly if two equivalent paths between the nodes exist and thus waste half the bandwidth. A link state protocol contains a full map of the network and can thus choose wisely between the paths or use both half of the time.

- *Fast, loopless convergence.* The distance vector protocol requires in the worst case the number of nodes minus one calculation step to calculate the algorithm, whereas the link state protocol only requires a flooding of all information and a local calculation of the optimal route.

- *Support for precise metrics and, if needed, multiple metrics.* It takes time to obtain exact information about a link; this causes the distance vector protocol to converge more slowly, when the metrics becomes more precise, and thus take more time to obtain the exact information.

- *Separate representation of external routes.* If no changes are made, a distance vector protocol automatically addresses the closest (default) gateway. If for some reason, however, another gateway needs to be selected, this gateway has to be added explicitly to the distance vector. In a link state protocol this could have been added quite easily.

3.4.3 Routing in Ad Hoc Networks

Cellular systems and Internet networks supporting mobile users rely on a fixed infrastructure, even if the end nodes can move. A base station can always reach all the mobile nodes in the cell without routing via a broadcast. In case of an ad hoc network:

- A fixed infrastructure is lacking and the topology of the network can quickly change.

- A destination node might be out of range of a source node transmitting packets.

- Each node must be able to forward data to other nodes;

These differences cause problems to dynamic routing algorithms designed for wired networks, which assume that the network topology does not change during the transmission. They would either react much too slowly or generate too many updates in the routing tables to reflect all changes in topology. The updating frequency of routing tables that is typical for wired network, 30 seconds, might be too low for ad hoc networks. Centralized approaches will not work in this case as it takes too long to collect the current status and distribute it. In presence of an unknown topology, the flooding approach is an alternative, which does not work when the load is high.

Moreover, most of the routing algorithms in wired networks rely on symmetric links in which routing information collected for one direction can be used for the other direction. In ad hoc networks, links can be asymmetric and the collected information for one link is not useful at all for the other one.

Interference among nodes creates new problems. Close by nodes that simultaneously forward transmissions might interfere and destroy each other. It can be observed that routing algorithms for this scenario cannot rely only on layer 3 knowledge. Information from lower layers, concerning connectivity or interference, can help to find a good path.

Suitable algorithms for ad hoc networks should also take care of the limited battery power of the routers that are basically nodes of the network. Some examples of routing algorithms used in ad hoc networks can be found in [11, 12].

3.4.4   Route Optimization in Mobile IP

One of the problems in Mobile IP version 4 is the asymmetric routing, also called *triangle routing*. The triangle routing is due to the fact that all packets going to the mobile first have to pass through the HA, while in the reverse direction packets are routed through regular IP routing. *Route optimization* [3, 13, 14] is the set of extensions to the basic Mobile IP protocol to allow more efficient routing procedures, so that the IP packets can be routed from a correspondent node to a mobile node without first passing through the HA.

The home agent sends a binding update message with the COA of the mobile node to all correspondent nodes that require them. This binding is then stored by the correspondent node and is used to tunnel its own IP packets directly to the care-of address indicated in that binding. Thus, passing through the HA is avoided. In the initiation phase, however, the IP packets sent by the correspondent node still use the triangle routing until the moment that the binding update message sent by the HA is received by the correspondent node.

Other extensions are provided to allow IP packets that are sent by a correspondent node with an out-of-date stored binding, or in transit, to be forwarded directly to the new COA of the MN.

The authentication mechanisms used in route optimization are the same as those used in the basic version of Mobile IP. This authentication generally relies on a mobility security association established in advance between the sender and receiver of such messages. The route optimization protocol operates in four steps:

1. A *binding warning* control message may be sent to the HA indicating that the correspondent node is unaware of the new COA of the mobile node.

2. A *binding request* message is sent by a correspondent node to the HA when it determines that its binding should be refreshed.

3. An authenticated *binding update* message is sent by the HA to those correspondent nodes that require them, containing the current COA of the mobile node.

4. When smooth handoffs occur, the mobile node transmits a binding update and has to be sure that the update has been received. Thereby, it can request a *binding acknowledgment* from the correspondent node.

The procedure of handoff in Mobile IPv4 is shown in Figure 3.6.

When a mobile node attempts at undertaking a handoff from one foreign domain to another, it sends a deregistration message to the previous foreign agent (e.g., FA1). The mobile node can send a deregistration message to FA1 or just make a handoff and let its connection with FA1 time out.

After the mobile enters a new foreign network, it waits for an agent advertisement from a FA. As soon as the mobile node receives the advertisement, it sends registration request to the home agent using the address of the new foreign agent (FA2) as care-of address. The HA processes the request and sends back a registration reply.

The signaling diagram for Mobile IPv4 handoff with binding update is shown in Figure 3.7.

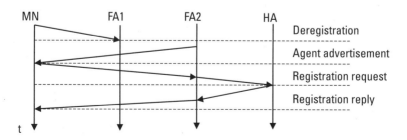

Figure 3.6 Mobile IP handoff signaling diagram.

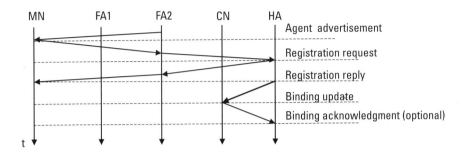

Figure 3.7 Mobile IP handoff with binding update.

3.5 IP QoS

There is a widespread consumer expectation that commercial IP-based mobile devices and fixed IP devices will need to provide QoS equal to that of cellular digital circuit-switched telephony. The support of QoS in any network requires the use of network resources, and the primary objective is to allocate and manage all dedicated bandwidth, control jittering, bound latency (required for real-time and interactive traffic), as well as meet data rate and reliability commitments.

The two current proposed QoS techniques over IntServ and DiffServ will be described in Sections 3.5.1 and 3.5.2, respectively. Section 3.5.3 is dedicated to the MPLS, which is a component of the IP QoS framework and applies to the two mentioned approaches. Finally, QoS issues related to Mobile IP and to heterogeneous networks are highlighted.

3.5.1 IntServ and RSVP

Integrated services architecture is described in detail in [15]. It was designed to provide a set of extensions to the best effort traffic delivery model in place in the Internet. In the model, in addition to the best-effort class of service, two other classes are defined:

1. Guaranteed service for applications requiring fixed delay bound;
2. Controlled-load service for applications requiring reliable and enhanced best-effort service.

No specific target loss and delay values are guaranteed.

According to this model, in order to deliver QoS, paths should be set up and resources reserved before data is transmitted. Traffic control mechanisms are fundamental building blocks of the IntServ architecture. Four mechanisms comprise the traffic-control functions:

1. *Packet scheduler,* which forwards packets based on their QoS class;

2. *Packet classifier,* which identifies within a host or router the required/ requested QoS level;

3. *Admission control,* which determines whether a flow can access the network with the requested QoS without affecting other established flows in the network;

4. *Resource reservation protocol,* which implements resource reservation in the end systems as well as in each router along the end-to-end flow path.

The ReSerVation Protocol (RSVP) [16] defines the signaling mechanism for the resource reservation. The RSVP signaling process is depicted in Figure 3.8.

The sender sends a *PATH* message to the receiver, specifying the traffic characteristics. Each router along the path forwards the PATH message to the next hop in the route. The receiver gets the PATH message and immediately responds with a *RESV (RESERVE)* message to request resources for the data flow. Each intermediate router on the way can accept or reject the request of the RESV message. If the request is accepted, the router allocates bandwidth and buffer space for the data flow, while installing the related *flow state* information. Otherwise, the router will send an error message to the receiver, thereby terminating the signaling process.

The scalability of RSVP in larger networks raises some concerns. The resource requirements such as computational processing and memory consumption for running RSVP on routers increase in direct proportion to the number of separate RSVP reservations, or sessions accommodated. Therefore, supporting a large number of RSVP reservations could introduce a significant negative impact on router performance. Given the current form of RSVP specification, multimedia applications run within smaller, private networks are the most likely to benefit from the deployment of RSVP.

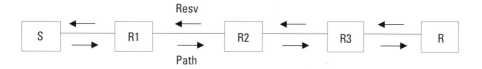

S = Sender
R = Receiver
R1, R2, R3 = Routers

Figure 3.8 RSVP operation.

3.5.2   DiffServ

The differentiated services scheme provides a less complex and more scalable solution. While the IntServ maintains per-flow states in each node, the DiffServ only discriminates among datagrams in different classes and not in individual flows. Consequently, the DiffServ approach allows different QoS levels to different classes of aggregated traffic flows, as opposed to individual flows, thus achieving scalability. Group of flows with similar requirements for resources can be aggregated into a single class. Each aggregated class is given a differentiated treatment within the network. To let customers receive differentiated services, they must have a service level agreement (SLA), an agreement between customer and service provider that specifies the service classes supported and the amount of traffic allowed in each class. The service provider should deliver the guaranteed level of service. To distinguish between the different service classes, DiffServ priority bit can be used. Unlike the case of IntServ, when using DiffServ, QoS guarantees are static; all traffic that passes a DiffServ capable router receives the required QoS. The only thing that needs to be done is to mark the type of service (TOS) or priority bit, which will be interpreted by the network routers. Within DiffServ the specification of how a packet should be handled (forwarding treatment) is called per-hop behavior (PHB). DiffServ requires mechanisms for queuing and scheduling the received packages according to the desired QoS. The commonly used FIFO queue does not offer service differentiation. Up to now, work on DiffServ at IETF has mainly focused on defining low-level forwarding mechanisms, while the design of the management plane remains a research issue.

3.5.3   MPLS

Multi-protocol label switching (MPLS) was originally designed as a means to speed-up IP routing, and now it is rapidly becoming the preferred vehicle to provide QoS over IP [17–19]. MPLS provides connection orientation based on IP routing and control protocols. It has been introduced as a new, standardized approach for integrating IP with ATM and is the recommended method for IP over ATM by the ITU [20]. The network layer could be any of the various network protocols in use today, such as IP, IPX, AppleTalk, thus, the significance of multiprotocol in the MPLS. However, because of the huge deployed base in the global Internet and the growing base of IP networks elsewhere, the first and foremost concern and application for this technology is to accommodate the IP protocol. Alternatively, MPLS can be implemented natively in ATM switching hardware, where the labels are substituted for (and situated in) the virtual path/virtual channel (VP/VC) identifiers.

MPLS forces traffic into specific, possibly nonoptimal routes, or label-switched paths (LSPs). In the MPLS network, routers are called label-switching routers (LSRs). Edge LSRs provide the interface between the external IP network and the LSP, while core LSRs provide transit services through the MPLS cloud. On the ingress, the edge LSR accepts IP packets and appends a fixed-length label between the network-layer header (Layer 3) and the link-layer header (Layer 2). Packets with common forwarding path requirements are assigned the same forward equivalence class (FEC). All packets that belong to a particular FEC and that travel from a particular node will follow the same path. Such a set of packets is called "stream." The stream or an FEC to which the packet is assigned is encoded in a short fixed length value known as a label. As is shown in Figure 3.9, the 20-bit label field is part of the MPLS "shim" packet header, which also contains a 3-bit experimental field (Exp), an indicator of additional labels (S), and TTL.

An MPLS node forwards packets based on the label value and not on the IP information. In the current IP forwarding method, referred to as *longest match*, a router references a routing table of variable-length prefixes and installs the longest, or most specific, prefix as the preference for subsequent forwarding mechanisms. The amount of time and computational resources required to make these decisions according to this forwarding method based on IP information is directly proportional to the number of prefixes and possible paths, which is very high in the global Internet and it is going to increase. In the MPLS node, where the forwarding is based on a set of fixed-length values, as opposed to the same number of variable-length values, it is computationally less intensive and, thus, takes less time. On the egress, an edge LSR terminates the LPS by removing MPLS labels and resorting to the normal IP forwarding.

QoS has several possibilities with MPLS. A seamless interworking between MPLS and DiffServ can be achieved by mapping DiffServ PHB to MPLS FEC [21]. Then the DiffServ class can be inferred from the label within the MPLS network. One possibility is a direct mapping of the 3 bits carried in the ToS precedence of the incoming packet headers and containing the class of service (CoS) information to the 3-bit Exp field in the MPLS header. Another possibility is that the label itself carries the CoS information. In this case, QoS mapping needs to be more defined. In both cases, mapping is handled by the ingress edge LSRs.

Moreover, MPLS is attractive for traffic engineering functions. Traffic engineering is the process of arranging how traffic flows through the network so

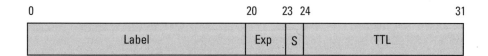

Figure 3.9 MPLS "shim" header.

that congestion caused by uneven network utilization, caused by current dynamic routing protocols, is avoided. Such a process can enhance service quality. Its attractiveness for traffic engineering can be attributed to the following factors:

- Capability to easily build explicit label-switched paths from one end of an MPLS network domain to another through manual administrative actions or through automated actions by the underlying protocols;
- MPLS allows for both traffic aggregation and de-aggregation using label stacking, whereas classical destination-based IP forwarding permits only aggregation.

Additionally, it is relatively easy to integrate MPLS with a "constraint-based routing" framework, which is an important tool for making the traffic engineering process automatic. Basically, constraint-based routing is used to compute routes that are subject to multiple constraints [22].

3.5.4   QoS in Mobile IP

RSVP and Mobile IP are not interoperable because of the following problems:

- States reservation is not possible over the *tunnel* between an HA and an FA. A PATH message that is sent by a correspondent node to the mobile node address is intercepted, encapsulated, and forwarded to the COA as any other IP datagram. According to the current RSVP specifications, both RSVP messages and reserved data flows are invisible to intermediate tunnel routers. Therefore, installing the *path state* [17] (and the corresponding *reservation state*) between a home and a foreign agent is not possible.
- However, if RSVP succeeds in installing reservation states over tunnels, the resulting RSVP performance over Mobile IP could be extremely inefficient.

There are several proposed solutions for QoS provisioning in Mobile IP. First, the proposal from the IETF Mobile IP Working Group focused on Mobile IPv4 protocol to resolve the RSVP operation over IP tunnels is discussed. Then, several alternative proposals from various research groups are outlined.

3.5.4.1   RSVP Operation over IP Tunnels

The specification of an IP tunneling mechanism is given in [23]. This is a standard proposed from the IETF Mobile IP Working Group for Mobile IPv4. This mechanism enables reservations across all IP-within-IP tunnels. The participation of a tunnel in the operation of an RSVP aware subnetwork (Figure 3.10) is accomplished through the following tunnel types:

1. The link does not provide QoS guarantees. This is referred to as a *best effort* or *type 1 tunnel.*
2. The link does not allocate resources to individual data flows, but it can provide QoS guarantees to aggregate flows. This type of tunnel is referred to as a *type 2 tunnel.*
3. The link can make reservations for individual end-to-end flows. The tunnel that participates in this operation is referred as *type 3 tunnel.*

IP tunnel mechanisms proposed in [23] can operate in all three tunnel types. As illustrated in Figure 3.10, *Rentry* represents the tunnel entry router, which encapsulates data into the tunnel. *Rexit* is the tunnel exit endpoint. In a best-effort tunnel, RSVP messages cross the link correctly. Also, the presence of the noncontrolled link has to be detected.

In order to guarantee QoS, the tunnel types 2 and 3 have to support a mapping between *end-to-end RSVP sessions* and *tunnel RSVP sessions.* For the tunnel type 2, a static mapping from an end-to-end RSVP session to an existing tunnel RSVP session can be achieved. For tunnel type 3, the mapping is dynamic, thereby creating a new tunnel RSVP session for each end-to-end

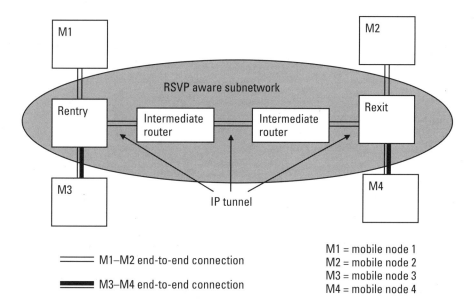

Figure 3.10 The RSVP operation over IP tunnels.

RSVP session. After a mapping between two tunnel sessions is done, it is necessary to coordinate the actions of the two RSVP sessions. End-to-end RSVP control messages, which have to be forwarded through a tunnel, are encapsulated like normal IP packets, so that the tunnel entry point is specified as source and the exit point is specified as destination.

3.5.4.2 Mobility Management in IP Networks Providing Real-Time Services

An effective way of including mobility in RSVP is proposed in [24]. It allows real-time communications setups between mobile and fixed hosts in TCP/IP networks. The approach is based on the "mobility notification method," which is an early notification of the current position of a mobile node to the sender. The home agent performs this notification and it is very similar to the route optimization in Mobile IP. The summary of the proposed method is as follows:

1. The correspondent node delivers a *Path* message to the receiver (mobile node).

2. When the mobile node is not connected to its home network, the home agent captures the RSVP Path message and replies to the correspondent node with a *PathChange* message containing the COA of the mobile node and its own address (MOBILITY_ NOTIFICATION Object), without tunneling the original Path message to the foreign agent.

3. The PathChange message is received by the correspondent node, and after that it caches the binding between mobile node home address and COA. Then, it sends a new Path message to the mobile node, tunneling it to the COA. Hereafter, reservation setup works in a traditional way, with the possible exception of tunnels usage.

The process of setting up a real-time communication between a correspondent node and a mobile node is illustrated in Figure 3.11.

The algorithm implies that RSVP has to be extended with a new message (PathChange) and a new object class (MOBILITY_NOTIFICATION).

The proposed method has simple, efficient, and limited impact on the actual RSVP proposal. Its disadvantages are related to the unsolved security issues (i.e., the PathChange messages from home agents to correspondent nodes should be authenticated).

3.5.4.3 An Architecture for QoS Guarantees and Routing in Mobile Networks

An architecture that provides QoS support in wireless networks is proposed in [25]. It is hierarchical and is based on the concept of the QoS regions and the routing regions. This approach solves the issue of efficient Mobile IP aware reservation mechanisms.

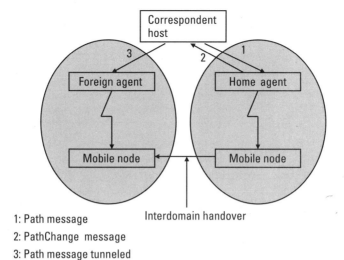

1: Path message
2: PathChange message
3: Path message tunneled

Figure 3.11 Reservation setup process.

QoS in Wireless/Mobile IP can be provided in several ways. One method is to reserve resources along the path(s) that the mobile might follow. This is an unnecessary waste of resources, since the reserved resources cannot be used. Another proposed method is the passive reservation scheme. The passive reservation guarantees that the reserved resources are not wasted when a mobile is not using them. Other applications can use the passive reservations until the moment required by the mobile for which the reservation is made. The passive reservations can be realized in two ways:

1. The sender makes passive reservations on all possible base stations in the vicinity.

2. Another designated node (e.g., base station) assists in creating the passive reservations on behalf of the sender.

The proposed architecture in [25] is capable of making passive reservations and balances the previously explained approaches.

A combination of Mobile IP and changes of the local routing table is used to form a hierarchical routing structure. According to the definition of routing domain:

• All the route changes within the routing domain are made by route table changes.

• The routing between routing domains is handled by using Mobile IP.

Furthermore, QoS domains are defined in the following manner:

- All passive reservations for mobility within a QoS domain are done by extending the path of the original reservation.
- Partial rerouting is used for all passive reservations between QoS domains.

The routing and QoS-based functionality of the proposed architecture is explained by the example illustrated in Figure 3.12.

- The CN is a sender. The MN is located in the subnet A and is communicating with the base station BSa. The correspondent node sends a request to the HA. The HA sends a reservation request to the FA (which also functions as routing-domain router). The correspondent node establishes a path to the FA using the mobile IP routing protocol, while the path from the FA to the mobile MN (through BSa) uses the routing table entries of the hosts in the same routing domain, which show the current location on mobile MN.
- The passive reservations are made locally while the mobile node remains in Subnet A. The QoS domain A router is requested by the BSa to invoke all its neighboring base stations in the passive reservation process. Two possible ways of making passive reservations are possible at this point.

1. When all neighboring base stations are in the same QoS domain, then the BSa can make passive reservations with all these neighbors.

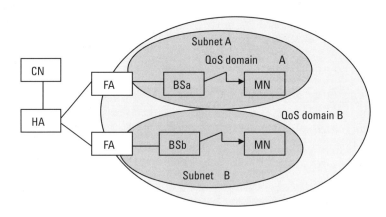

Figure 3.12 Architecture for QoS guarantees and routing in mobile networks.

2. When some of the neighboring base stations are in a different QoS domain, then the BSa makes passive reservation with the base stations in the same QoS domain. The QoS domain router makes passive reservations with the base stations in the neighboring QoS domain.

- If the mobile moves into another subnet B (i.e., subnet B, managed by BSb), then two states are possible, either the reservations between BSa and BSb are activated or the reservations between the QoS domain router and BSb are activated, depending on which QoS domain BSb resides.

- In case the roaming is local (i.e., within the same routing domain), then the routing tables in the hosts in this domain are changed to indicate that the mobile is in the location of the BSb.

- If the mobile node roams to another routing domain, then the current routing-domain router (i.e., FA) informs the HA about the roaming and the HA chooses another QoS domain router to the other subnet as the FA.

A modified version of RSVP is used in the proposed approach. The modified RSVP has the following changes:

1. Passive reservation messages should be incorporated;
2. New QoS parameters that are specific to mobile environment, such as:

    - *Loss profiles:* In case of overloading, it gives an application the opportunity to choose between distributed loss or bursty loss.
    - *Probability of seamless communication:* It defines the nature of breaks in the service during handoff.
    - *Rate reduction factor:* It is used in situations when there is no possibility for requested passive reservations to be provided and consequently, it starts the procedure of renegotiations on a fraction of the resources.

3. The required TCP flow can use the passive reservations made between base stations.

3.5.4.4 A Signaling Protocol for Integrated Services Internet with Mobile Hosts

A framework to reduce the RSVP signaling overhead in mobile environment is introduced in [26]. This framework proposes an extended reservation model to enhance utilization of scarce wireless bandwidth. A mobile node has to make an advance resource reservation along the data flow paths at the locations, which it may visit during the lifetime of its connections.

Three classes of reservations are introduced:

1. *Conventional reservation* reserves bandwidth along the data flow path from sender to current location on wired link and the wireless bandwidth in the current cell.

2. *Predictive reservation* reserves bandwidth along the multicast tree from the source to the neighboring cells, surrounding the current cell of the mobile node.

3. *Temporary reservation* temporarily uses the inactive bandwidth reserved by other flows in the current cell.

There are no foreign agents in this architecture. A device called *mobile proxy* replaces the foreign agent entity. It makes various reservations along the data path from the current cell and the surrounding cells towards the sender. The mobile proxy at the current location of a mobile node is called *current mobile proxy*, while the mobile proxies at the cells surrounding the current cell are called *neighbor mobile proxies*. Mobile proxies are very similar to home agents and foreign agents in Mobile IP. They process the mobile related RSVP messages and maintain the mobile soft state for mobile nodes.

This approach supports mobility by extending the RSVP model based on IP multicast. RSVP messages and IP packets are delivered to a mobile node using the IP multicast routing. The mobility of a node is modeled as a transition to multicast group membership. The multicast tree is modified dynamically every time a mobile node is moving from one cell to another.

The current mobile proxy propagates the conventional reservation message toward the sender along the multicast tree. The predictive reservation messages from neighbor mobile proxies are forwarded in the same way. If these reservations are successful, the data packets can be transmitted over the conventional reservation link and resources on the predictive reservation link and neighboring cells are reserved for this data flow. When the mobile receiver is moving from one location to another, the flow of data packets can be switched to the new route as quickly as possible. The multicast tree is modified dynamically, each time a mobile node is roaming to a neighboring cell.

This method requires a larger background processing and a higher consumption of bandwidth on wired links. However, it eliminates the need for rerouting the data path during the handoff.

3.5.5 IP QoS Among Heterogeneous Networks

The currently proposed QoS techniques, such as IntServ, DiffServ, and MPLS, are either nonscalable, too immature, or both to enforce and to manage end-to-end QoS throughout an IP-based heterogeneous network. The issue is how to effectively combine the existing protocols together across heterogeneous networks. Figure 3.13 illustrates a simple QoS reference model for various real-time IP communications, which covers call setup, QoS reservation, policy, authorization, and payments [27].

Key elements in the hierarchical model include the access network, backbone network, and clearinghouse (i.e., centralized QoS management unit).

- *Access network:* An IP network to which users directly connect their hosts/clients for IP connectivity. The access network is part of a single administrative domain, such as networks, government, and educational organizations. Access network elements are responsible for QoS management, which supports client (IP device) QoS. Client QoS can be invoked by the user either via an interface mechanism or by running a QoS-aware application.

- *Backbone network:* One or more backbones networks may be between two or more access networks. The backbone network in the model has no knowledge of individual microflows, such as phone calls between parties connected to access networks.

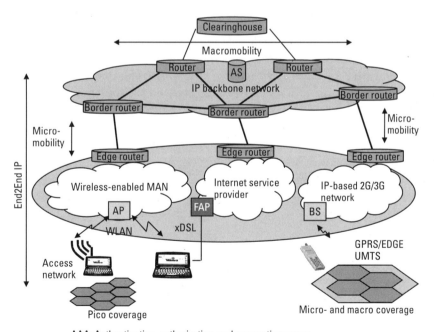

AAA: Authentication, authorization, and accounting server
AS: Application server   AP: Access point   FAP: Fixed access point
BS: Base station/node B

Figure 3.13 Reference model for QoS support in IP-based heterogeneous networks.

- *Clearinghouse:* Given the large number of access networks belonging to different administrative domains, it is not possible to have service-level specifications (SLS) between all domains on the Internet. Clearing-houses can facilitate the authorization and logging or accounting between domains for premium services, such as QoS. Present SLS are static in nature, although there is interest in signaling for dynamic delivery of QoS between service providers, such as in the case of bandwidth-broker-mediated services.

Strict priority for real-time traffic, such as voice, facilitates simple and robust QoS implementation in a private IP network, where policy control and individual specific accounting are not required. However, extension of the strict priority to real-time traffic between the different domains across the Internet would prevent service providers from exercising policy control and accounting, which would make services very costly and reduce the incentive for service providers to deploy QoS mechanisms in their networks. Furthermore, the model of strict priority service without policy and accounting may be under pressure when applied to high-bandwidth applications, such as video on demand. Economic trade-offs between the subscription charge and the guaranteed QoS level will be required.

One of the difficulties in adding QoS to the backbone network is to implement mechanisms whose scaling properties are suitable for high-capacity infrastructure, where resources may be shared by enormous numbers of simultaneous traffic flows. The statistical properties of highly aggregated traffic flows generally make the maintenance of per-microflow state unnecessary to provide adequate end-to-end performance, even for applications such as telephony. Although individual networks may vary greatly, a static allocation of resource per traffic class is generally sufficient to provide high statistical assurance of adequate end-to-end service, provided that:

- A large number of microflows can share resources during periods of peak load.

- No single microflow is large enough to utilize a large percentage of a potentially congested resource.

- The total amount of traffic in the network receiving priority over best-effort traffic is relatively low.

The approach proposed in [27] uses standard end-to-end RSVP in access networks and DiffServ with MPLS in the backbone networks. DiffServ within MPLS networks is a good candidate to provide QoS in backbone networks because services are based on the per-hop model and the aggregate forwarding policies are preallocated in the LSR for each service type. RSVP messages from the access network are tunneled through the backbone network [28, 29]. MPLS facilitates scalability because it aggregates flows, ensuring individual end-to-end QoS guarantees without the need to maintain awareness of individual flows on each path segment.

Backbone networks are not visible to end users, except when end-to-end QoS cannot be provided. The access network does not need to understand the QoS mechanism. The only common element between the backbone and the access networks is the SLS, which specifies the DSCPs for various classes of traffic passing the backbone network.

3.6 Security Issues

The basic goals of security management are:

- *Data confidentiality:* Information is available for authorized parties only;
- *Data integrity:* Unauthorized parties cannot modify information;
- *Authentication:* Assurance of the identity of a peer entity or the origin of information;
- *Non-repudiation:* An entity cannot later on falsely deny a valid transaction;
- *Access control:* Unauthorized persons cannot access resources;
- *Availability and correct functioning of services:* Information is available for authorized users when they need it [30–32].

Current digital mobile networks provide a robust set of security facilities to protect communications across the air interface. The main security services offered by GSM are (1) access control and authentication, which includes the authentication of a valid user for the SIM and the subscriber authentication; (2) confidentiality of user and signaling data through encryption; and (3) anonymity.

Future networks will need to support a larger range of security services. First of all, the needs of users of these multimedia services are potentially very different from those of voice users of existing networks. One important issue is the end-to-end integrity and confidentiality. Existing networks do not support integrity. GSM offers confidentiality only between MS and BTS, but not end-to-end or within the whole fixed GSM/telephone network. New mechanisms of user identity confidentiality based on both public key and conventional cryptographic techniques have been examined. End-to-end confidentiality raises political as well as technical problems. Law enforcement agencies should be able to access to certain communication paths, when warrant exists, but this access should be controlled because of the civil liberties issues. A trusted third parties (TTPs)-based scheme for warranted access has been developed, offering considerable advantages over its competitors [33].

Moreover, security management in current mobile systems is not standardized. Although the management security requirements are clear, the exact way in which user key information is generated, stored, and accessed is left to network operators (NO) and equipment providers to arrange. This can make security service provision costly for all concerned, since every NO may arrange security management differently. In future heterogeneous networks, possibly operating in a more deregulated environment than the present one, standardized support for security management will be a very important feature. Without any standards, the required cooperation between the large numbers of competing NOs and service providers could become too complex.

Several standardized security architectures exist for securing IP communications including TCP, UDP, as well as application-specific transport protocols: IPSec at IP level, Secure Sockets Layer (SSL) at the transport layer, proxy and authentication, authorization, and accounting (AAA) servers at application level. In the rest of the section, some of these IP security mechanisms will be presented, in particular, IPSec and the AAA framework. The extensions of these security mechanisms to Mobile IP are currently under consideration for standardization within the IETF. In Section 3.6.3, a comparison of the basic Mobile IP security models is given. Then, various solutions to all security issues, basically proposed from the IETF Working Groups, are discussed. Few solutions, proposed from non-IETF research groups, are also considered, since they can be integrated into the Mobile IP architecture.

3.6.1 IPSec

The IP security architecture IPSec has been standardized by the IETF [34, 35]. With IPSec, security mechanisms are directly applied to IP packets. As the contentionless IP does not create any context between two communicating entities, the concept of security association (SA) is introduced by IPSec architecture. An SA consists of three parameters: destination address, cryptography protocol, and an identifier to separate multiple SAs with the same destination host but using the same cryptography protocol. Two SAs should be defined, because the SAs are unidirectional. Two extensions are defined to protect IP packets:

1. The *authentication header* (AH) is used to protect the authenticity and integrity of an IP packet with a keyed cryptographic hash value.

2. The *encapsulating security payload* (ESP) that transports encrypted IP packets, ensuring confidentiality and optionally authenticity as well as integrity of the packet payload. Contrary to AH, the outer IP header is not protected.

Authentication mechanisms work on a packet-per-packet basis (connectionless) and provide both authentication and integrity check; optional reply protection can be used. The AH mechanism authenticates as much of the packet as possible; however, some parameters, including TOS, flags, TTL, and the header checksum, change during the connection and, hence, cannot be authenticated. Modifications made to the IP packets are more complicated in the case of ESP than in the case of AH. Besides the insertion of an extra header, a trailer and some authentication data are also included.

Two modes are defined for both ESP and AH: *transport* and *tunnel* mode. The first refers to the mode where the original packet is taken and then modified. In the tunnel mode, tunneling is applied and a new packet is constructed.

3.6.2 AAA

Authentication involves validating the identity of end users, prior to permitting them network access. *Authorization* defines what rights and services the end user is allowed once network access is granted. This might include providing an IP address, invoking a filter to determine which applications or protocols are supported, and so on. *Accounting* provides the methodology for collecting information about the end user resource consumption, which can then be processed for billing, auditing, and capacity-planning purposes. AAA essentially defines a framework for coordinating these disciplines across multiple network technologies and platforms.

Within the fixed Internet, several authentication and authorization approaches are already used for dial-up computers. The best-known and most deployed AAA protocol is RADIUS, acronym for Remote Access Dialing User Service [36]. RADIUS is a protocol that carries authentication, authorization, and configuration information between a client, referred to as a network access server (NAS), and an authentication server. Another AAA protocol is DIAMETER [37]. Similarly to RADIUS, DIAMETER provides authentication, authorization, and accounting. In addition, it provides policy control and resource control. DIAMETER and RADIUS protocols can coexist and interwork.

3.6.3 Security Issues in Mobile IP

3.6.3.1 Mobile IP Extensions with AAA

The basic Mobile IP model that employs AAA is proposed in [38]. This model is shown in Figure 3.14.

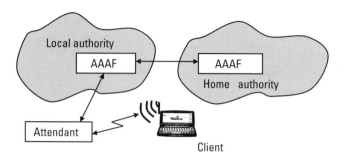

Figure 3.14 The basic Mobile IP/AAA model.

In traditional IP networks, there is a frequent situation when a client that belongs to one administrative domain (home domain) requires resources from another domain (foreign domain). In the foreign domain, there is a designated agent (called the *attendant*) that attends to the client request. It requires some credentials (permit) from the client, which have to be authenticated before it permits an access to the resources.

A local authority (AAAF) in the same foreign domain is invoked to obtain proof that the client has legal credentials. The attendant and the AAAF have security associations, which are used for secure local information transaction.

It is likely that AAAF will not be able to authorize the client. However, AAAF is able to negotiate the client authorization with another external authority, that is, the home domain AAA server (AAAH) for the client. The local and external authorities set up enough security associations and access control so that they can negotiate the authorization. The authorization is successful if a secure authentication of the client's credentials is achieved. After a successful authentication, the attendant provides the client with the requested resources.

There are several specific security associations assumed in the basic Mobile IP/AAA model, as illustrated in Figure 3.15.

This is a significant difference between the basic mobile IP security model in RFC 2002 [1] and this model. Unlike RFC 2002, the AAA model assumes that the client has a security association SA_1 with the AAAH. In RFC 2002, the mobile node has a security association with the home agent, which is a simple routing agent. Several requirements to this model can be identified:

- A security association is required between each local attendant and the local AAA server (AAAF).

- There must be security associations between the local authority and the external authorities, which can authenticate customer credentials.

- While the local authority contacts the appropriate external authority, the attendant has to keep state for pending customer requests.

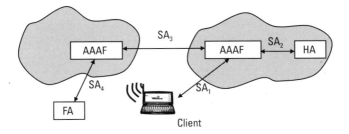

Figure 3.15 The security associations in the basic Mobile IP/AAA model.

- The mobile node must be able to provide complete unforgeable credentials without the prior requirement to interact with its home domain. Otherwise, it will not get the requested service.

- The attendant has to protect against replay attacks.

- The attendant or AAAF must not be allowed to learn any (secret) information, which might compromise the credentials.

The AAA server has additional general tasks for Mobile IP:

- It should authorize the mobile node to use Mobile IP and certain specific services.

- It has to initiate and enable the authentication for Mobile IP registration.

- It has to distribute keys to the MN and mobility agents.

Each repetition of the AAA key distribution process can cause lengthy delays between registrations. Therefore, any security associations distributed by the AAA server should have sufficiently large lifetime to avoid such delays. The consequences of these registration delays are dropped packets and noticeable disruptions in service [1]. The lifetime of the key between the MN and the HA determine the time interval before the MN has to initiate a new key distribution from the AAA servers.

3.6.3.2 Reverse Tunneling for Mobile IP

The reverse tunneling for Mobile IP technique, proposed in [39], offers a solution to the Mobile IPv4 ingress filtering. It specifies a reverse tunnel. The starting point of the tunnel is the mobile node COA and the endpoint is the home agent. After a mobile node moves to a foreign network, it detects foreign agents by listening to agent advertisements. The advertisements contain information that the foreign agent supports a reverse tunnel. It requests this service when it registers through the selected foreign agent. Afterward, the mobile node selects a packet delivery style. It selects either the *direct* or the *encapsulating delivery style*.

In the direct delivery style, the mobile sends a packet directly to the foreign agent without encapsulation. The foreign agent intercepts the packets and it tunnels them to the home agent. The source address of the outer header of the packets is the mobile node COA. It is considered that the home agent can also support the reverse tunneling procedure.

In the encapsulating delivery style, the mobile node encapsulates all its packets to the foreign agent. The foreign agent deencapsulates the packets and retunnels them to the home agent, using the foreign agent COA as the entry point of this new tunnel.

3.6.3.3   Mobile IP Extension for Private Internets Support (MPN)

An extension to the Mobile IPv4 basic protocol, which includes security issue, is proposed in [40]. This extension enables mobility that covers multiple routing domains. It offers solutions to the ingress filtering. The basic Mobile IP protocol uses IP-within-IP as a default encapsulation protocol. This document proposes the GRE protocol as the default tunneling protocol. The protocol uses a combination of the source route entry (SRE) and the *address family* option, thereby specifying intermediate destinations. Thus, it avoids the encapsulation procedure used in Mobile IPv4, and the tunneling procedure can be applied on different both private and public networks. In case of MPN the source route entry will include any intermediate mobility agent (in private or public domains) along the tunnel route and the tunnel endpoint.

3.6.3.4   Use of IPSec in Mobile IP

A method for applying IPSec [41] onto the IP-within-IP encapsulation used by Mobile IP to redirect IP packets to and from the mobile nodes is proposed in [42]. It solves the authentication in Mobile IP. The proposed scheme includes:

- A mechanism for negotiating the use of IPSec protection on selected Mobile IP redirection tunnels;
- A procedure for establishing these IPSec protected tunnels;
- The formats of tunneled packets in either full IP-within-IP or minimal IP-within-IP encapsulations.

This scheme provides authentication and privacy services to Mobile IP redirection traffic. In this way, Mobile IP traffics are protected against passive and active attacks, gaining a possibility to pass through security gateways.

3.6.3.5 Registration Keys for Route Optimization

A method for providing a security association between a mobile node and a foreign agent at the moment the mobile node registers with this foreign agent is introduced in [43]. It proposes a solution for the authentication and the authorization. The binding update messages (see [13]) that might change the routing of IP datagrams to the mobile node require authentication. A security relationship between the sender and receiver of such messages is established in advance. Whenever the mobile node moves to a foreign network, such security association between the mobile node and the new foreign agent is difficult to be realized in advance. It is necessary for the foreign agent to have such a security association to be able to handle the eventual binding updates that it may receive from the mobile node. These binding updates provide a mechanism for accomplishing smooth handoffs between a previous foreign agent and a new foreign agent. The operations that are performed during the smooth handoffs should be secure. It means that the both foreign agents (the previous and the new foreign agent) have to be sure that they are getting authentic handoff information. In [43], the special messages accomplishing authentication, are used together with the Mobile IP registration request and the registration reply messages. The exact identity of the foreign agent is not critical to the process of establishing a registration key. Several methods are specified in [43]. These methods are hereinafter outlined:

- The foreign agent and mobile node share a security association. This can be used to secure the previous foreign agent notification without need to establish a registration key.

- When a home agent and a foreign agent share a security association, the home agent can choose the new registration key.

- If the mobile node can transfer key information between foreign agents that trust each other, it can use the same key utilized with its previous foreign agent.

- If the foreign agent has a public key, it will require the home agent to supply a registration key.

- When the mobile node includes its public key in its registration request, then the foreign agent can choose the new registration key.

- The foreign agent and the mobile node can execute a Diffie-Hellman key exchange protocol [44] as part of the registration protocol.

Once the registration key is established, the smooth handoff method can be used. Other methods of establishing keys may become available in the future.

3.6.3.6   Mobile IP Challenge/Response

The specification in [45] defines a number of extensions, for the Mobile IPv4 agent advertisements and the registration request. They enable a foreign agent to use a mechanism called *challenge/response* to authenticate the mobile node. This mechanism provides solutions to the authentication and the authorization. The challenge is a random value of at least 128 bits and is used to compute an authentication procedure. A trusted association between a foreign agent and a home agent is created by using a mechanism of verification infrastructure (see Figure 3.16).

After receiving a challenge response from the mobile node, the foreign agent passes it to the entity called, *Verification and Key Management Infrastructure*, and waits for a registration reply. The foreign agent accepts the registration from the Verification and key Management Infrastructure only if the reply is positive. If the reply is negative, then the foreign agent assumes that the challenge has not passed the verification.

3.6.3.7   Mobile IP Public Key–Based Authentication

In [46] an authentication extension to the basic Mobile IP protocol is provided. It defines the way mobile nodes and mobility agents may use public key or secret key based authentication by digital signatures. This extension solves the authentication and the encryption key distribution. The secure scaleable authentication (SSA) approach is applied in this authentication extension. It takes advantage of a few reserved fields in the existing Mobile IP message definitions. The increased functionality of SSA enables modifications to the authentication extension used in the Mobile IPv4 protocol in order to accommodate different authentication types and different sizes of authenticators (digital signatures). Also, the use of either IP address or host name can be used to identify mobile nodes and mobility agents.

3.6.3.8   New Secure Minimal Public Key–Based Authentication

The base Mobile IP is based on the use of secret keys with manual key distribution for authentication of its control messages. This approach is inappropriate for wide scale application, since it definitely leads to scalability problems in key management. Moreover, even though authenticated registration messages and replay protection are already deployed, the current registration protocol suffers

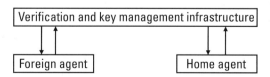

Figure 3.16 Verification infrastructure.

from a possible replay attack. The proposed solution in [46] is an additional attempt to provide scalable authentication mechanism based on public key cryptography. Its drawback, however, is the heavy requirement on the mobile node to perform demanding certificate-based public key cryptography operations.

The protocol proposed in [47] employs only minimal use of public key cryptography. The design principles and assumptions in this protocol are as follows:

- Minimization of computing power requirement as well as administration cost imposed on MN. It achieves scalability and non-repudiation, while keeping public-key operations minimal.

- It does not require any additional message exchange except the original Mobile IP control messages. This gives a mobile node a flexibility to choose the desired authentication scheme.

The mechanism for certificate retrieval and validation is not addressed in this document. Only a public key infrastructure (PKI) that is used by home agents and foreign agents is assumed. The protocol proposed in [47] solves the authentication and the non-repudiation.

3.6.3.9 Mobile IP Network Address Identifier Extension

A network address identifier (NAI) extension to the Mobile IPv4 registration request [1] message is specified in [48]. Solutions to the issue of authentication and the issue of authorization are provided. With this extension, a mobile node may even not have a home address, and still it can authenticate itself and be authorized to connect itself to the foreign access network. It introduces a new function named the Home Domain Allocation Agency (HDAA) (see Figure 3.17), which can dynamically assign a home address to the mobile node.

When some message contains the mobile node NAI extension, it may have the home address zero (0) field in the registration request. The foreign agent must hence use the NAI as a replacement for its pending registration request

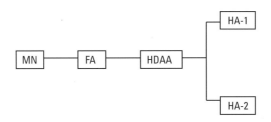

Figure 3.17 HDAA.

records. The HDAA shown in Figure 3.17 receives messages from foreign agents and assigns a home address within the home domain. The HDAA does not perform any Mobile IP processing on the registration request, but is simply forwarding the request to the HA that can handle it.

3.6.3.10   NAI Resolution for Wireless Networks

The method proposed in [49] offers an opportunity to match the wireless cellular subscriber identification to/from the NAI during the wireless registration and authentication process. It also provides a solution to the issue of using one single subscription for all service types. Through minor modifications, this solution can also be applied for Mobile IPv6. The NAI-enabled wireless cellular service provider owns the cellular service for subscriber A (SUB A) as depicted in Figure 3.18.

The operation of the mechanism is described in the following steps:

1. SUB A powers on his second or third generation cellular mobile node. Automatically, the mobile node attempts a wireless registration. The mobile node is registered and identified by its mobile identification number (MIN).

2. The wireless access network receives the wireless registration message, which matches a NAI based on the MIN sent by the cellular mobile node. After that, the wireless access network sends an appropriate registration message to its NAI enabled home network.

3. After receiving the registration message, the NAI-enabled home network registers and authenticates the wireless SUB A. Afterwards, it sends an appropriate registration response back to the wireless access network.

4. After receiving the registration response, the wireless access network sends an appropriate wireless registration return result to SUB A cellular mobile node.

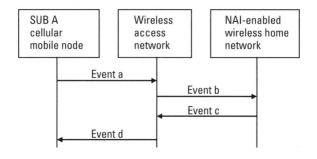

Figure 3.18 NAI resolution scenario.

3.6.3.11 DIAMETER Mobile IP Extensions

Interdomain (different ISPs) authentication and authorization are provided in [50], which specifies extensions to DIAMETER [37]. These extensions solve the issue of authentication and the issue of authorization. In addition, it specifies the assignment of mobile node home address, assignment of home agent, as well as a key distribution approach. AAA DIAMETER servers are located in different ISP domains and accomplish these Mobile IPv4 solutions using associations (communications) among them.

3.6.3.12 Firewall Support for Mobile IP

The issue of firewall support in Mobile IP is solved with the specification given in [51]. It can also be applied for Mobile IPv6, probably with minor modifications. Two methods can be used to provide firewall support for mobile nodes. The first method is based on application proxying (relaying), where each mobile node establishes a TCP session (established to port 1080) to exchange UDP traffic with the firewall. This method is accomplished using the SOCKS protocol (version 5) [52]. The second method is based on IP security, where the traffic from the mobile node to the firewall could be encrypted and authenticated using a session-less IP security mechanism like SKIP [53].

A Firewall-Aware Transparent Internet Mobility Architecture

The aim of the Firewall-Aware Transparent Internet Mobility Architecture (FATIMA) [54] is to define an architecture that integrates the advantages of the following approaches: Mobile IP, Cellular IP, HAWAII, and Hierarchical Mobile IP. At the same time, it avoids their disadvantages and provides a solid base for adding new features (dynamically assigned home addresses or quality of service mechanisms) in a consistent and straightforward manner. FATIMA introduces new network entities:

- *FATIMA gateway* is the central entity that supports mobility inside a network. Actually, it is an internal component of the network firewall.
- *Home/foreign proxies* (HAP/FAP) replace all home/foreign agents. They convey most of their functionalities to the FATIMA gateway.
- *Routing agents* (RA) are set up hierarchically in order to improve scalability in large networks. They are used to connect the FATIMA gateway with each home/foreign agent proxy in the network.

An overview of this architecture is presented in Figure 3.19. By using ESP tunnels, all control and data traffic between entities of the mobility infrastructure are authenticated. Moreover, mutual authentication of mobile nodes and the foreign network infrastructure is ensured. The architecture significantly

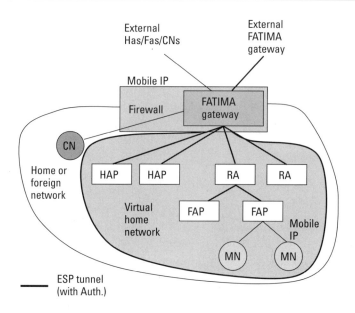

Figure 3.19 FATIMA network entities.

improves the efficiency of local handoffs (micromobility) together with route optimization. It can be extended with mechanisms that provide complete mobility support. FATIMA provides seamless support for private addresses, which is a requirement in today's nonacademic IP networks. Also, dynamically assigned home addresses can be supported.

3.7 Evolution of Mobile IP

Mobile IPv4 is inefficient to meet the requirements of the evolving communication scenario for several reasons.

Macrombility

Open issues in Mobile IPv4 related to macromobility management are:

- *Asymmetric (triangle) routing* (see Section 3.4).
- *Inefficient direct routing:* The routing procedure in Mobile IPv4, with respect to the number of hops or end-to-end delay is inefficient.
- *Inefficient home agent notification:* When a mobile node handoffs from one network to another, it has to notify the home agent about that. This operation is inefficient in Mobile IPv4.
- *Inefficient binding deregistration:* In Mobile IPv4, if a mobile node moves to a new foreign agent, then the previous foreign agent could not

release the resources occupied by the mobile node. The previous foreign agent must wait until a binding registration lifetime expires.

Micromobility

Mobile IPv4 is not concerned with micromobility issues. Since it is expected that in near future, Mobile IP will be a core wireless IP architecture interconnecting different wireless networks, the interoperability between macromobility and micromobility issues is very important. There are several open issues in this context:

- *Local management of micromobility events:* In micromobility dimension, handoffs occur very frequently. Therefore the handoff procedures should be managed as much as possible locally.
- *Seamless intradomain handoff:* After intradomain handoff, the IP data stored into the previous entities (e.g., base stations) should be transferred to the new BS.
- *Mobility router crossings in an Intranet:* During intradomain handoff, the router crossings should be as much as possible avoided.

3.7.1   Mobile IPv6

The IETF began work in 1994 on IPv6, a proposed standard designed to address various problems with Mobile IPv4. Before IPv5, some of these problems had already been addressed by the IPv5 standard. Nevertheless, the IPv.5 standard has been skipped as network operators had already adopted most of the modifications proposed for the standard.

It is supposed to replace Mobile IPv4. One of the main advantages of IPv6 over IPv4 is the higher number of IP addresses. IPv4's 32-bit addressing scheme can support a theoretical maximum of 4.29 billion IP addresses. Due to the operational inefficiency, however, useful IP addresses are about 200 million. IPv6 offers a 128-bit addressing scheme that permits about useful IP addresses.

Other major differences between the Mobile IPv4 and the Mobile IPv6 [55] are:

- Mobile IPv6 supports the mechanism of route optimization. This feature is already an integral part of the Mobile IPv6 protocol. In Mobile IPv4 the route optimization feature is just a set of extensions that may not be supported by all IP nodes.
- Mobile IPv6 specifies a new feature that allows mobile nodes and Mobile IP to coexist efficiently with routers that perform "ingress filtering" [56]. The packets sent by a mobile node can pass normally through ingress filtering firewalls. This is possible due to the fact that the COA

is used as the source address in each packet IP header. Also, the mobile node home address is carried in the packet in a *home address destination option* [55]. In this way, the use of the COA in the packet is transparent above the IP layer.

- The use of the COA as the source address in each packet IP header simplifies the routing of multicast packets sent by a mobile node. Thus, the tunneling of the multicast packets to its home agent will no longer be necessary in Mobile IPv6. The home address can be still used and compatible with multicast routing that is based in part on the packet source address.

- Neighbor discovery [57] and address autoconfiguration [58] enable the functionality of the foreign agents instead of using special routers. The foreign agents are not required any more in Mobile IPv6. Thus, the issue of mobility router crossings in an intranet is resolved.

- The Mobile IPv6 uses IPsec for all security requirements [e.g., sender authentication, data integrity protection, and replay protection for binding updates (which serve the role of both registration and route optimization in Mobile IPv4)]. Mobile IPv4 is based on its own security mechanisms for each function, based on statically configured mobility security associations.

- Mobile IPv6 provides a mechanism for supporting bidirectional (i.e., packets that the router sends are reaching the mobile node, and packets that the mobile node sends are reaching the router) confirmation of a mobile node ability to communicate with its default router in its current location. This bidirectional confirmation can be used to detect the "black hole" situation, where the link to the router does not work equally well in both directions. Unlike Mobile IPv6, Mobile IPv4 does not support bidirectional confirmation. Only the forward direction (packets from the router are reaching the mobile node) is confirmed, and therefore the black hole situation may not be detected.

- In Mobile IPv6, the correspondent node sends packets to a mobile node while it is away from its home network using an IPv6 routing header rather than IP encapsulation, whereas Mobile IPv4 must use encapsulation for all packets. In this way RSVP operation in Mobile IP is enabled and also the problem of ingress filtering is partially solved. In Mobile IPv6, however, the home agents are allowed to use encapsulation and tunnel the packets to the mobile node.

- In Mobile IPv6, the home agent intercepts the packets, which arrive at the home network and are destined for a mobile node that is away from home, using IPv6 neighbor discovery [57] rather than ARP [9] as is used in Mobile IPv4.

- IPv6 encapsulation (and the routing header) removes the need in Mobile IPv6 to manage "tunnel soft state," which was required in Mobile IPv4 due to limitations in ICMP error procedure for IPv4. In Mobile IPv4, an ICMP error message that is created due to a failure of delivering an IP packet to the COA, will be returned to the home network, but will not contain the IP address of the original source of the tunneled IP packet. This is solved in the home agent by storing the tunneling information (i.e., which IP packets have been tunneled to which COA, called *tunneling soft state*).

- Mobile IPv6 defines a new procedure, called *anycast*. Using this feature, the dynamic home agent discovery mechanism returns one single reply to the mobile node, rather than the corresponding Mobile IPv4 mechanism that used IPv4 directed broadcast and returned a separate reply from each home agent on the home network. The Mobile IPv6 mechanism is more efficient and more reliable. In this way, only one packet needs to be replied to the mobile node.

- In Mobile IPv6, an advertisement interval option on router advertisements (equivalent to agent advertisements in Mobile IPv4) is defined, which allows a mobile node to decide for itself how many router advertisements (agent advertisements) it is ready to miss before declaring its current router unreachable.

- The IPv6 destination options permit all Mobile IPv6 control traffic to be piggybacked on any existing IPv6 packets. Mobile IPv4 and its route optimization extensions require separate UDP packets for each control message.

3.7.2 Macro/Micromobility Extensions to Mobile IP

Several protocols and frameworks have been proposed to extend Mobile IP to better support micro- and macromobility in next-generation wireless cellular environments.

3.7.2.1 Fast and Scalable Handoffs for Wireless Internetworks

This is an extension to Mobile IP [59] that uses hierarchical FAs to handle macromobility. This architecture assumes BSs to be network routers; for that reason, it is not compatible with current cellular architectures, where BSs are simply layer 2 forwarding devices. Moreover, deploying a hierarchy of FAs imposes complex operational and security issues (especially in a commercial multiprovider environment) and requires multiple layers of packet processing over the data transport path. The presence of multiple layers of mobility-supporting agents also significantly increases the possibility of communication failure, since it does not exploit the inherent robustness of Internet routing protocols.

3.7.2.2   Mobile-IP Local Registration with Hierarchical Foreign Agents

This protocol is an IETF proposal [60], and it solves the issues of triangle routing and inefficient direct routing. The basic mechanism of this architecture deploys hierarchical FAs for seamless mobility within a domain. During the COA discovery procedure multiple foreign agents are advertised using the agent advertisement message. The COA registration will be provided for the foreign agent that is the lowest common foreign agent ancestor at the two points of attachment of interest. The requirement for hierarchical agents in Internet mobility architecture remains an open issue. Even though it does not appear to be a critical consideration in the immediate future, it is possible that hierarchical mobility management will become more attractive as the IP security infrastructure matures and deployment of mobile multimedia terminals gets much larger.

3.7.2.3   TeleMIP

TeleMIP stands for Telecommunication Enhanced Mobile IP [61]. This architecture faces the problem of inefficient home agent notification. It achieves smaller handoff latency by localizing the scope of most location update messages within an administrative domain or a geographical region. TeleMIP introduces a new logical entity, called the *mobility agent* (MA), which provides a mobile node with a stable point of attachment in a foreign network. While the MA is functionally similar to conventional foreign agents, it is located at a higher level in the network hierarchy than the subnet-specific FAs. Location updates for intradomain mobility are localized only up to the MA. Global location updates are necessary only when the mobile changes the administrative domains. The TeleMIP allows efficient use of public address space by permitting the use of private addresses for handling macromobility. Reduction of the frequency of global update messages overcomes several drawbacks of existing protocols, such as large latencies in location updates, higher probability of loss of binding update messages, and loss of in-flight packets, and thus provides better mobility support for real-time services and applications. The dynamic creation of mobility agents (in TeleMIP) permits the use of load balancing schemes for the efficient management of network resources. Its drawback is the potential nonoptimal routing within the domain.

3.7.2.4   Wireless IP Network Architecture by TR45.6

Another framework for IP-based mobility management was developed by the Telecommunications Industry Association (TIA) Standards Subcommittee TR45.6 [62] to target 3G cellular wireless systems. This architecture is consistent with the requirements set by the ITU for IMT-2000. Therein, solutions are provided to the issue of inefficient direct routing. The framework uses Mobile IP with fast rerouting for global mobility. For macromobility, the scheme proposes the use of dynamic HAs, which reside in the serving network and are dynamically assigned by the visited AAA server. The DHA allows the roaming user to gain service with a local access service provider while avoiding unnecessarily long routing. The architecture defines a new node called a *packet data-serving node* (PDSN) (which contains the FA), and uses VLR/home location register (HLR) (ANSI-41 or GSM-MAP) authentication and authorization information for the access network. The mobile node is identified by an NAI [63] in the visiting or foreign network. Within the registration process, an MN sends the registration message to the FA, which in turn interacts with an AAA server residing in that network or uses the broker network for authentication with the home network.

3.7.2.5 Micromobility Extensions to Mobile IP

Due to the fact that the basic Mobile IP protocol [1] is only concerned with the macromobility management, some other solutions are required to enhance the Mobile IP functionality to support micromobility. The following sections overview the results of several current research activities in this area.

Wireless Network Extension Using Mobile IP

This micromobility management framework [64] is combined with Mobile IP. Thus, it provides solutions to the issue of local management of micromobility events. The development of this scheme is realized in the Motorola iDEN architecture. Micromobility events should be managed more efficiently than macromobility events, because they can happen with relatively high frequency. Therefore, the procedures and participants are being kept as local as possible. The micromobility procedures are managed by a data gateway, thus achieving the previous condition. The macromobility between iDEN subnetworks and other subnetworks is accomplished by implementing Mobile IPv4 in the foreign agent and home agent (Figure 3.20).

Handoff-Aware Wireless Access Internet Infrastructure

The Handoff-Aware Wireless Access Internet Infrastructure (HAWAII) [65] proposes a method for using a separate binding protocol to handle micro- and macromobility. For global mobility, it uses Mobile IP. Using this architecture, a solution to the issue of local management of micromobility events is provided. It uses a two-layer hierarchy for mobility management. When the MN moves into a foreign domain, it is assigned a collocated COA from that domain, and the MN retains its COA unchanged while moving within the foreign domain. Thus, the movement of the MN within a domain is transparent to the HA. This

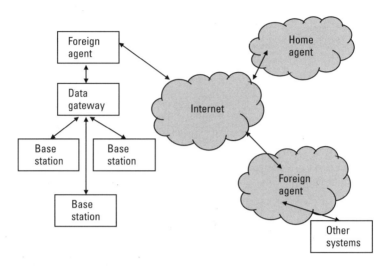

Figure 3.20 Mobile IP wireless network extension.

protocol uses path setup messages to establish and update host-based routing entries for MNs in some specific routers within the domain; other routers not in the path are kept in the dark about the MN new COA. When a CN sends packets to a roaming user, it uses the MN home IP address. The HA intercepts the packets and sends the encapsulated packet to the MN's current border router. The border or root router decapsulates and again encapsulates the packet to forward it to either the intermediate router or BS, which decapsulates the packet and finally delivers it to the MN (Figure 3.21).

Cellular IP

Cellular IP [66, 67] proposes an alternative method to provide mobility and handoff support in a cellular network, which consists of interconnected cellular IP nodes. Solutions to issues of local management of micromobility events and seamless intradomain handover are provided. This protocol uses Mobile IP for global mobility. It is very similar to the host-based routing paradigm of HAWAII. Specifically, Cellular IP provides local mobility (e.g., between BSs in a cellular network) (Figure 3.22).

The architecture uses the home IP address as a unique node identifier, since MN addresses have no location significance inside a cellular IP network. When an MN enters a Cellular IP network, it communicates the local gateway (GW) address to its HA as the COA. Nodes outside the Cellular IP network do not require any enhancements to communicate with nodes inside the network. When a CN sends packets to a roaming user, it uses the MN's home IP address. As in conventional Mobile IP, the HA intercepts the packets and sends the encapsulated packet to the MN current GW. The GW decapsulates the packet

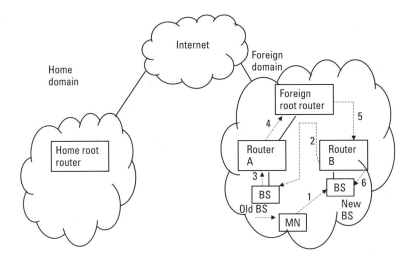

Figure 3.21 HAWAII architecture.

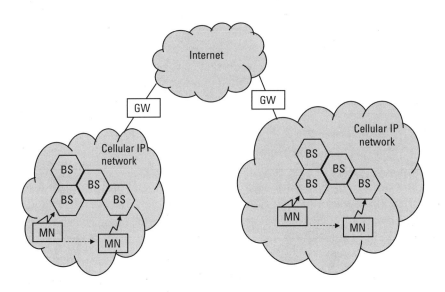

Figure 3.22 Cellular IP.

and forwards it to the MN's home address using a node-specific route. Thus, the nodes sending or receiving packets to/from the MN remain unaware of the node location inside the Cellular IP network.

HAWAII and Cellular IP have very similar concepts. However, there is a major difference between them. In the HAWAII protocol, most of the intelligence is in the network part, while in Cellular IP most of the intelligence is in the mobile node. Therefore, Cellular IP is not optimal for management of the security and QoS. However, the network equipment is simpler and therefore cheaper.

3.7.3 RSVP Support for Mobile IP Version 6

The solution for the RSVP operation over IP tunnels is provided for Mobile IPv6 as well [68]. The specification in the draft proposes three solutions:

1. The first solution requires modifications in RSVP at both mobile and correspondent nodes, so that they have to be aware of Mobile IPv6 addressing.

2. In order to enhance the performance and make handoffs smooth and seamless, optional *triggers/objects* are added to RSVP messages. The RSVP PATH messages are triggered on bindings updates and home address objects that are contained in RSVP RESV messages. Thus, intermediate routers are enabled to recognize connections and to use resources even when the COA changes.

3. A *flow extension* mechanism is provided. This mechanism is able to extend the existing RSVP flows (i.e., flow_ids) that are applied on typical IP routers to the new Mobile IP router. It is combined with a simultaneous binding option that has to be applied for the roaming mobile node. The mobile node receives packets on both previous and current COA.

To make Mobile IPv6 and RSVP interoperable, the minimal solution (1) is a requirement [68]. This requires the modification and the interfacing of the RSVP daemon and Mobile IP binding cache at both CN and MN. The latter two solutions (2 or 3) can provide uninterrupted operation since they support fast reestablishment or preservation of resource reservations when mobile nodes move. Table 3.1 presents a qualitative comparison of the latest two approaches.

It should be noted that triggers/objects is a quick solution with low complexity, which is able to provide sufficiently good service (e.g., for mobile IP phones). The flow extension approach is a little more complex but has the advantage of faster deployment. In multiprovider environments, where the whole end-to-end path could not be controlled, a solution that modifies only CNs, access network routers, and MNs has a big advantage.

3.8   Conclusions

Mobile IP has been proposed for network level mobility support. In cellular systems, like GSM and UMTS, mobility within the system is required to be handled by the system itself.

Table 3.1

RSVP Support for Mobile IP Version 6: Qualitative Comparison of Approaches 2 and 3

Criteria	Triggers/Objects	Flow Extension
Changes to CN	Yes (required for minimal solution)	Yes (required for minimal solution)
Changes to intermediate routers	Yes (RSVP MIP object extension and reuse of flow resources)	No
Changes to MIP-router	No (forwarding of late packets is also an option here)	Yes (binding update interception, flow forwarding)
Changes to MN	Yes	Yes
Changes to HA	No	No
Supports multicast delivery	Yes	Yes
Bandwidth efficiency	Yes	Yes (it is assumed efficient overprovisioning in the access network)
End-to-end delay	Always shortest path (but reestablishment of resources requires a round-trip)	Slightly increased delay
Lossless HO	Yes (with forwarding of late packets)	Yes
HO delay	Roundtrip	Faster
Implementation complexity	Moderate	Higher

From an IP network point of view mobility is handled by the link layer and does not affect the IP layer as long as the user stays in the same network type. However, when a user connects to the network via different network types or through different operators network, the roaming mechanism is needed. Mobile IP version 4 represents a simple and scalable global mobility solution but is not appropriate in support of fast and seamless handover. Mobile IP version 6 offers a more robust, secure and scalable set of features with a significantly increased address space.

What can be envisaged about the future of IPv6? Because the technology is so new, the protocol evolution could take various directions. While the marketing strategy of many IPv6 proponents was focused on convincing wireline ISPs about adopting the protocol, the focus has now moved towards wireless technology. IPv6 may be most widely deployed in mobile phones, PDAs, and other wireless terminals in the near future. Europe's Third Generation Partnership Project, which is working on next generation wireless-network technology, has specified IPv6 as the standard addressing scheme for mobile IP multimedia. So far it has received support only in some geographical areas, like Asia. Equipment and OS vendors have just begun supporting IPv6. There is a relatively little support for IPv6, particularly in North America, from the Internet service providers and network administrators. There are contrasting opinions on the reason of such a little support. Industry observers believe that it is partly due to the cost and effort required to migrate to IPv6, as well as the protocol shortcomings and a perception that standard advantages are not so necessary. Others claim that the widespread adoption will take place in years, because IPv6 represents a major change in Internet technology.

Open issues in Mobile IPv6 include the following:

- Triangle routing and inefficient direct routing;
- Local management of micromobility events;
- Seamless intradomain handoff;
- Efficient Mobile IP aware reservation mechanisms;
- RSVP reservations on Mobile IP triangle route situations;
- The use of one single subscription for all service types;
- Firewall support.

Resource reservation is necessary for providing guaranteed end-to-end performance for multimedia applications. The issues of QoS and resource reservation have been studied in great detail, even if the first and key step in resource reservation is routing. A communication can be undertaken if a path between sender and receiver, which meets the specific requirements the application has set, can be found. Resources can, hence, not be reserved unless the routing protocol can find a suitable path. Current routing algorithms used in IP networks are transparent to any particular QoS that different flows could require. Routing decisions are made without referring to the QoS of the flow. This means that flows are often routed over paths that are unable to support their requirements, while alternate paths with sufficient resources exist. Finding the shortest path to the destination is not enough anymore. New routing algorithms have to be developed with the aim of finding a path in the network that satisfies the given requirements [69]. However, the proposed QoS routing protocols should require minimal changes to the existing ones. QoS routing is a very important research issue. The new complexity of routing is a problem for designers and implementers, and a trade-off must be made between the optimal solution and its complexity.

Although important research areas can be found in the framework of the provision of QoS in IP networks, critical elements already exist for implementing QoS-enabled IP networks. What is needed is a practical architecture able to combine all the established and underdevelopment technologies in a more and more effective heterogeneous environment.

References

[1] Perkins, C., (ed.), "IP Mobility Support," RFC2002, proposed standard, IETF Mobile IP Working Group, Oct. 1996.

[2] Schooler, E., et al., "SIP: Session Initiation Protocol," IETF RFC 2543, Mar. 1999.

[3] Perkins, C., "Mobile IP," *IEEE Communications Magazine*, May 1997.

[4] Deering, S. E., (ed.), "ICMP Router Discovery Messages," RFC 1256, Sept.1991.

[5] Perkins, C., "Minimal Encapsulation Within IP," RFC2004, October 1996.

[6] Hanks, S., et al., "Generic Routing Encapsulation over IPv4 Networks," RFC 1701, Oct.1994.

[7] Jacobson, V., "Compressing TCP/IP Headers for Low-Speed Serial Links," RFC 1144, Feb. 1990.

[8] Perkins, C., "IP Encapsulation Within IP," RFC 2003, Oct.1996.

[9] Plummer, D., "An Ethernet Address Resolution Protocol: Or Converting Network Protocol Addresses to 48.bit Ethernet Eddresses for Transmission on Ethernet Hardware," RFC 826, Nov. 1982.

[10] Huitema, C., *Routing in the Internet*, Engelwood Cliffs, NJ: Prentice Hall, 1995.

[11] Perkins, C., and P. Bhagwat, "Highly Dynamic Destination-Sequenced Distance Vector Routing (DSDV) for Mobile Computers," *Proc. ACM SIFCOMM 1996*.

[12] Johnson, D. B., and C. Perkins, "Dynamic Source Routing in Ad Hoc Wireless Networks," in *Mobile Computing* (Imieliski/Korth), Kluwer Academic Publishers, 1996.

[13] Perkins, C., and B. Johnson, "Route Optimization in Mobile IP," Internet draft, draft-ietf-mobileip-optim-10.txt, work in progress, Nov. 2000.

[14] Perkins, C., "Mobile Networking Through Mobile IP," *IEEE Internet Computing*, 1998.

[15] Braden, R., D., Clark, and S. Shenker, "Integrated Services in the Internet Architecture: An Overview," RFC 1633, June 1994.

[16] Braden, R., et al., "Resource Reservation Protocol (RSVP)—Version 12 Functional Specification," Aug. 12, 1996. Available at http://www.ietf.org/html.charters/intserv-charter.htm.

[17]   Ors, T., and C. Rosenberg, "Providing IP QoS over GEO Satellite Systems Using MPLS," *International Journal of Satellite Communications*, Vol. 19, 2001, pp. 443–461.

[18]   Finenberg, V., "A Practical Architecture for Implementing End-to-End QoS in an IP Network," *IEEE Communications Magazine*, Jan. 2002, pp. 122–130.

[19]   Ferguson, P., and G. Huston, *Quality of Service-Delivering QoS on the Internet and in Corporate Networks*, New York: John Wiley & Sons, 1998.

[20]   ITU Draft Recommendation I.ipatm, "IP over ATM," Sept. 1999.

[21]   Le Faucher, F., "MPLS Support of Differiantiated Services," Internet-draft, IETF MPLS Working Group, March 2000.

[22]   Jamoussi, B., (ed.), "Constraint-Based LSP Setup Using LD," Internet-draft, IETF MPLS Working Group, Sept. 1999.

[23]   Terzis, A., et al., "RSVP Operation over IP Tunnels," RFC 2746, Jan. 2000.

[24]   Andreoli, G., et al., "Mobility Management in IP Networks Providing Real-Time Services," *Proc., Annual International Conference on Universal Personal Communications*, 1996, pp. 774–777.

[25]   Mahadevan, I., and M. Sivalingham, "An Architecture for QoS Guarantees and Routing in Wireless/Mobile Networks," *ACM Intl. Workshop on Wireless and Mobile Multimedia*, 1998.

[26]   Chen, W.-T., and L.-C. Huang, "RSVP Mobility Support: A Signaling Protocol for Integrated Services Internet with Mobile Hosts," *INFOCOM 2000*, Vol. 3, pp. 1283–1292.

[27]   Alam, M., R. Prasad, and J. R. Farserotu, "Quality of Service Among IP-Based Heterogeneous Networks," *IEEE Personal Communications*, Dec. 2001.

[28]   Bernet, Y., "The Complemntary Roles of RSVP and Differentiated Services in the Full-Service QoS Network," *IEEE Communication Magazine*, Feb. 2000.

[29]   Rouhana, N., and E. Horlait, "Differentiated Services and Integrated Services Use of MPLS," *Proc. IEEE ISCC2000*, pp. 194–199.

[30]   Ford, W., *Computer Communications Security: Principles, Standard Protocols and Techniques*, Englewood Cliffs, NJ: Prentice Hall, 1994.

[31]   Kaufman, C., R. Perlman, M. Speciner, Network Security—Private Communication in a Public World, Prentice Hall, 1995.

[32]   Stalling, W., *Cryptography and Network Security: Principle and Practice*, Second Edition, Englewood Cliffs, NJ: Prentice Hall, 1998.

[33]   Jefferies, N., C. Mitchell, and M. Walker, "A Proposed Architecture for Trusted Third Party Services," In E. Dawson and J. Golic, *Cryptography: Policy and Algorithms*, Springer-Verlag LNCS 1029, 1996, pp. 98–104.

[34]   Kent, S., and R. Atkinson, "IP Authentication Header," RFC 2402, Nov. 1998.

[35]   Thayer, R., N. Doraswamy, and R. Glenn, "IP Security Document Roadmap," RFC 2411, Nov. 1998.

[36]   Rigney, C., et al., "Remote Authentication Dial In User Service (RADIUS)," RFC 2138, April 1997.

[37]   Calhoun, R., "DIAMETER," Internet draft, draft-calhoun-diameter-07.txt, work in progress, Nov. 1998.

[38]   Perkins, C., "Mobile IP Joins Forces with AAA," *IEEE Personal Communications,* Aug. 2000.

[39]   Montenegro, G., "Reverse Tunnelling for Mobile IP," RFC 3024, Jan. 2001.

[40]   Teo, W., and Y. Li, "Mobile IP Extension for Private Internet Support (MPN)," Internet draft, draft-teoyli-mobileip-mvpn-02.txt, work in progress, 1999.

[41]   Kent, S., and R. Atkinson, "Security Architecture for the Internet Protocol," Internet draft, draft-ietf-ipsec-arch-sec-02.txt, work in progress, Nov. 1997.

[42]   Zao, J., M. Condell, "Use of IPSec in Mobile IP," Internet draft, draft-ietf-mobileip-ipsec-use-00.txt, work in progress, Nov. 1997.

[43]   Perkins, C., and D. Johnson, "Registration Keys for Route Optimization," Internet draft, draft-ietf-mobileip-regkey-03.txt, work in progress, July 2000.

[44]   Diffie, W., and M. Hellman, M., "New Directions in Cryptography," *IEEE Transactions on Information Theory,* Vol. 22, Nov. 1976, pp. 644–654.

[45]   Perkins, C., P. R. Calhoun, "Mobile IP Challenge/Response Extensions," RFC 3012, Nov. 2000.

[46]   Jacobs, S., "Mobile IP Key Based Authentication," Internet draft, draft-jacobs-mobileip-pki-auth-02.txt, work in progress, March 1999.

[47]   Sufatrio, K. Y. L., "Mobile IP Registration Protocol: A Security Attack and New Secure Minimal Public-Key Based Authentication," SPAN '99, June 1999, pp. 364–369.

[48]   Calhoun, P. R., C. Perkins, "Mobile IP Network Address Identifier Extension," Internet draft, draft-ietf-mobileip-mn-nai-01.txt, work in progress, May 1999.

[49]   Aravamudhan, L., M. R. O'Brien, and B. Patil, "NAI Resolution for Wireless Networks," Internet draft, draft-ietf-mobileip-nai-wn-00.txt, work in progress, Feb. 1999.

[50]   Calhoun, P., and A. C. Rubens, "DIAMETER Reliable Transport Extensions," Internet draft, draft-calhoun-diameter-mobileip-01.txt, work in progress, Feb. 1999.

[51]   Montenegro, G., and Gupta, V., "Firewall Support for Mobile IP," Internet draft, draft-montenegro-firewall-sup-03.txt, work in progress, Sept. 1996.

[52]   Leech, M., et al., "SOCKS Protocol Version 5," RFC 1928, March 1926.

[53]   Aziz, A., and M. Patterson, "Design and Implementation of SKIP," available on-line at http://skip.incog.com/inet-95.ps, 1995.

[54]   Mink, S., et al., "FATIMA: A Firewall-Aware Transparent Internet Mobility Architecture," *ISCC'2000,* July 2000, pp. 172–179.

[55]   Johnson, D. B., and C. Perkins, "Mobility Support in IPv6," Internet draft, draft-ietf-mobileip-ipv6-13.txt, work in progress, Nov. 2000.

[56]   Ferguson, P., and D. Senie, "Network Ingress Filtering: Defeating Denial of Service Attacks Which Employ IP Source Address Spoofing," RFC 2267, Jan. 1998.

[57]   Narten, T., E. Nordmark, and W. A. Simpson, "Neighbour Discovery for IP version 6 (IPv6)," RFC 1970, Aug. 1996.

[58]   Thomson, S., and T. Narten, "IPv6 Stateless Address Autoconfiguration" RFC 1971, Aug. 1996.

[59]   Caceres, R., and V. N. Padmanabhan, "Fast and Scalable Handoffs for Wireless Internetworks," *Proc. MOBICOM '96*, ACM, Aug. 1996, pp. 76–82.

[60]   Perkins, C., "Mobile IP Local Registration with Hierarchical Foreign Agents," IETF Internet draft, work in progress, Feb. 1996.

[61]   Das, S., A. Misra, and P. Agrawal, "TeleMIP: Telecommunications-Enhanced Mobile IP Architecture for Fast Intradomain Mobility," *IEEE Personal Communications*, Vol. 7, Aug. 2000, pp. 50–58.

[62]   Hiller, T., (ed.), "Wireless IP Network Architecture Based on IETF Protocols," Ballot v. PN-4286, TIA/TR45, June 1999.

[63]   Perkins, C., and P. R. Calhoun, "Mobile IP Network Access Identifier Extension for IPv4," Internet draft, draft-ietf-mobileip-mn-nai-07.txt, work in progress, July 1999

[64]   Geiger, R. L., J. D. Solomon, and K. J. Crisler, "Wireless Network Extension Using Mobile IP," *IEEE Micro*, Vol. 17, No. 6, 1997, pp. 63–68.

[65]   La Porta, T., R. Ramjee, and L. Li, "IP Micro-Mobility Support Using HAWAII," Internet draft, draft-ietf-mobileip-hawaii-00.txt, work in progress, June 1999.

[66]   Valko, A. G., "Cellular IP: A New Approach to Internet Host Mobility," *Comp. Commun. Rev.*, Jan. 1999, pp. 50–65.

[67]   Wan, C.-Y., et al., "Cellular IP," Internet draft, draft-valko-cellularip-01.txt, work in progress, Oct. 1999.

[68]   Fankhauser, G., S. Hadjiefthymiades, and N. Nikaein, "RSVP Support for Mobile IP Version 6 in Wireless Environments," Internet draft, draft-fhns-rsvp-support-in-mipv6-00.txt, Nov. 1998.

[69]   Gosh, D., V. Sarangan, and R. Acharya, "Quality-of-Service Routing in IP Networks," *IEEE Transactions on Multimedia*, Vol. 3, No. 2, June 2001, pp. 200–208.

4

TCP over Wireless Links

4.1 Introduction

The Transport Control Protocol (TCP)/IP and UDP/IP protocol suites form the basis of the Internet. Most Internet applications today (i.e., Web browsing, Telnet, FTP), and also those likely in the near future, rely on the TCP transport layer protocol [1].

TCP protocol has been designed to perform well over wired and fixed networks. Significant performance degradation is observed over paths that include wireless hops and mobile hosts. In recent years, a great effort has been paid to improve TCP performance over different wireless environments. Several approaches have been proposed but no final consensus has been reached.

This chapter presents some of the proposed solutions and the ongoing research effort to face this key issue for future wireless networks. Section 4.2 is an overview of the main mechanisms within TCP that play an important role in using TCP over wireless links. IETF workgroups have recommended a set of configuration options. Some of the main configuration options that can be found in all modern versions of TCP are presented in Section 4.3. These recommendations are necessary to solve some of the limitations of standard TCP over wireless links, but other problems such as the error-prone characteristics of these links are not addressed effectively. The effects of noncongestion-related packet losses can be alleviated by splitting the connection into two or more parts at an access point to the wireless link. *Splitting TCP* and *Snooping TCP*, two possible approaches to split the connection, are also described in Section 4.3.

All these TCP enhancements require understanding and modifying TCP protocol behavior to account for wireless hops. Another approach consists of introducing protocols at the data link layer or lower layers that makes the channel appear more reliable to the TCP protocol, in such a way that most of the packet losses are congestion-related. The data link layer of some 2.5G and 3G technologies such as GPRS, W-CDMA, and cdma1x [2, 3] provides automatic repeat request (ARQ) error recovery mechanisms over the radio link. Basic ARQ mechanisms are recalled in Section 4.4 and their interactions with the TCP protocol are highlighted. Improved link layer mechanisms that exploit some knowledge about the channel where they operate are presented in Section 4.5. The identification of innovative link layer mechanisms to improve the end-to-end performance is an interesting research area and it is a part of the new framework in the wireless networks design, as discussed in Section 4.5. Conclusions are drawn in Section 4.6.

4.2 Standard TCP/IP Protocol

TCP is a connection-oriented transport layer protocol that provides:

- Reliable data delivery;
- In-sequence delivery of data:
- Reordering out-of-order data;
- Discarding duplicate data.
- End-to-end flow control.

Applications do not all require or all benefit from the reliable service provided by TCP. Applications where reliability is not critical or with strict transmission delays constraints, such as telephony over IP or audio and video streaming, are carried by the UDP protocol. UDP only provides multiplexing and error detection whereas reliability and in-order delivery are handled at the application layer.

TCP/IP is a surprisingly complex protocol suite. Several books are fully dedicated to its operation mechanisms. A detailed description of TCP is provided in [1]. The aim of this section is to outline those TCP mechanisms that most directly affect TCP performance in a wireless environment, rather than summarize all TCP/IP features. Therefore, the end-to-end flow and congestion control mechanisms are first introduced, showing also the implications of wireless networks features (channel characteristics, mobility) on the end-to-end performance of a protocol that has been designed and optimized for wired and fixed networks.

4.2.1 Sliding Window Mechanism

In order to achieve reliability, two hosts exchanging data through a TCP protocol have to exchange information on the state [i.e., confirmation messages (acknowledgments) that indicate whether a previous segment has been successfully received]. The sliding window mechanism renders this confirmation process more efficient as it allows the acknowledgment of a group of in-sequence received segments instead of one segment at a time. An example of the sliding window mechanism is shown in Figure 4.1.

In the example a transmission window of six segments is assumed. The transmission window, which measures the amount of unacknowledged data that the sender can have in transit to the receiver, is a key parameter for all the mechanisms described in this section. The window size, in bytes, is exchanged in each TCP segment. It is contained in a 16-bit field of the TCP segment

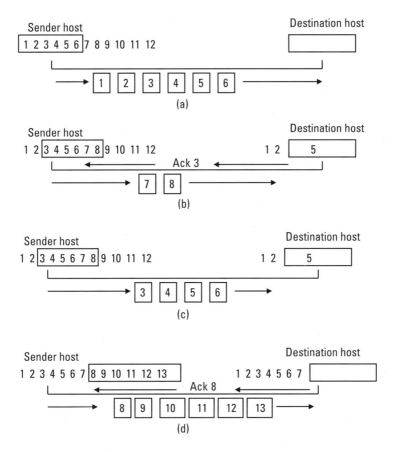

Figure 4.1 Sliding window mechanism: (a) The transmitter uses a sliding window of 6 segments. (b) Segments 1, 2, and 5 are received; the ACK of the reception of segments 1 and 2 is sent; and the sliding window shifts of two segments, thus allowing the transmission of segments 7 and 8. (c) Segments from 3 to 6 are retransmitted when the retransmission time-out expires. (d) The ACK of the reception of segments 1–7 is sent out and segments 8–13 can be transmitted.

header, so that it can change during the transmission. While the sender transmits the six segments, a retransmission time out is set for each segment, indicating when the segment has to be resent, in case it is not acknowledged. In fact, TCP has no negative-acknowledgment (NACK) mechanism (i.e., the receiver does not tell directly to the sender which segment is missing, but continues to acknowledge all in-sequence segments up to the missing one). The sender notices the missing acknowledgment for the lost segment when the retransmission time out (RTO) expires. In the example of Figure 4.1, segments 1, 2, and 5 only are received [Figure 4.1(b)]. Therefore, the receiver sends an acknowledgment that contains number 3, corresponding to the missing segment. The sliding window can now move on two positions and segments 7 and 8 can be sent. When the timer of the unacknowledged segments (3–6) expires, they are retransmitted [Figure 4.1(c)]. The RTO is now set to a value that is double of the previous value. When all segments up to 7 are received, an ACK is sent back to the receiver indicating that 8 is the next expected segment. The transmission window now shifts in order to transmit segments from 8 to 13 [Figure 4.1(d)]. RTOs are set for each segment and the process is repeated.

4.2.2   End-to-End Flow and Congestion Control

End-to-end flow and congestion control mechanisms should ensure data transmission at a rate consistent with the capacity of both the receiver and the intermediate links in the network path. If a link is idle, a TCP connection is expected to be able to fill the available bandwidth. If a link is shared with other users, each TCP connection is expected to get a reasonable share of the bandwidth. Flow and congestion control are basically achieved by controlling the transmission window size. The receiver advertises a window to the sender, which it is related to the amount of available buffer space at the receiver for the connection. The sender establishes a congestion window (CWDN), according to its assessment of perceived network congestion. Network congestion occurs whenever, in some network nodes, packet buffers are filled and the router cannot forward packets fast enough, because the sum of the input rates of packets destined to one output link is higher than the output link capacity. In this case, the router must drop some packets and a dropped packet is lost for the transmission. In wired networks with fixed end-systems, the main reason for losing packets is network congestion. When a packet loss is detected, by missing the acknowledgment for the lost packet, the TCP sender tries to mitigate congestion by slowing down the transmission rate and, hence, by reducing the transmission window, which is determined by:

$$current\ window = min(CWND, advertised\ window)$$

All other TCP connections experience the same congestion and do exactly the same. Therefore, the congestion is soon resolved. Nowadays this congestion control mechanism is one of the main reasons for the survival and stability of the Internet.

4.2.2.1 Slow Start and Congestion Avoidance

TCP uses an algorithm called *slow start* to probe the path and, hence, learn how much bandwidth is available. At the beginning, the transmission window is set to 1 [i.e., the sender sends a TCP segment and waits for an acknowledgment (ACK)]. Upon reception of an acknowledgment, the congestion window is doubled. If TCP acknowledges every other received segment, the slow start algorithm sends effectively 50% more data every round trip. This is called the exponential growth of the congestion window in the slow start mechanism. If the connection is about to use all the available bandwidth, it is too dangerous to double the transmission window. In fact, this could cause network congestion, thus degrading also the performance of other links. Therefore, a slow start threshold (SSTHRESH) is defined, which is initialized to the maximum window size. As soon as the congestion window reaches this threshold, the value of the receiver advertised window, or as soon as a congestion occurs, the slow start phase is terminated and the *congestion avoidance* phase starts. In this phase, the further increase of the transmission window is linear with the round-trip time (see Figure 4.2).

During this phase, if network congestion is assumed, because one or more segments have been lost, the sender sets SSTHRESH to half of the value of the current CWND and the CWND itself is set back to one segment. The window will start again an exponential growth, up to the SSTHRESH value and then, as described above, a linear growth.

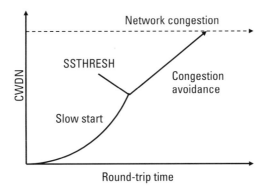

Figure 4.2 CWND versus RTT for slow start and congestion avoidance algorithms.

4.2.3   Implications of Large Bandwidth-Delay Product

The performance of flow and congestion control mechanisms are strongly affected by the value of the so-called Bandwidth*Delay (BD) product. The delay in the definition is the round trip time (RTT), and the bandwidth is the maximum bandwidth of the slowest link in the path. In this definition, bandwidth is interchangeable with bit rate. The BD product determines how much unacknowledged data volume a TCP sender should transmit into the network to fully utilize the link capacity. In the 1970s, when TCP was being developed, a typical long link across the United States was characterized by a BD product of 1.8 kB. Today, in 3G networks the BD product ranges between 8 and 50 kB. Assuming a 200-ms RTT and a service that offers 384 Kbps, the delay*bandwidth product is 76.8 kbits (9.6 kB), which is larger than the default 8 kB buffer space used by many TCP implementations. For a 2-Mbps service, the BD product is 50 kB. Networks that include these "long thin pipes" links are called "long thin networks" (LTNs) [4]. Networks that include satellite links are examples of "long fat networks" (LFNs or "elephants") [5–7]. They are "long" networks because their RTT is quite high (for example, 250 ms one way for a geosynchronous satellite), and they are "fat" in the sense that they may have high bandwidth. Satellite networks may often have a BD product above 64 kB. Not all satellite links fall within the LFN regime. In particular, round-trip times in a low-earth orbiting (LEO) satellite network may be as little as a few milliseconds (and never extend beyond 160 to 200 ms).

Standard TCP is inefficient over networks characterized by a BD product higher than default buffer space, basically because these networks might never experience the full available bandwidth for the connection [8].

A limitation to the full bandwidth exploitation in LFNs is the need of large buffers at the receiver, in order to handle large amount of data in a single window. This represents a problem for the satellite link since the provision of large buffers (i.e., big memories) on board is not an easy task.

In large BD systems, moreover, the probing algorithm in the slow start phase could take a long time to get up to speed (i.e., to start to use a good share of the available bandwidth). The time required to get up to speed can be estimated by the following expression [5]:

$$T_{slowstart} = RTT\left(1 + \log_{1.5}\left(\frac{BD}{l}\right)\right) \qquad (4.1)$$

where l denotes the average segment length. For instance, in a 155-Mbps GEO link with a 0.5s two-way RTT, assuming 1-kB datagrams, the slow start phase will take 11.3s. During this period most of the link bandwidth will be wasted. Furthermore, 20 MB of data could be sent during this period, and most of the typical Internet applications, like Web browsing, might complete the entire transfer during the slow start phase. A similar problem occurs during the congestion avoidance phase. The linear probing algorithm characterized by a rate increase of *1/BD*, might take quite a long time to reach an optimal steady-state transfer rate.

4.2.4 Implications of Link with Errors and Mobility

As previously described, congestion window and RTOs are the parameters that TCP exploits to adapt its use of the network capacity based on the feedback by the TCP receiver. The CWND is determined from an estimate of the available bandwidth, whereas RTOs are set from an estimate of the overall path delay. These estimates are performed on assumptions that are valid in wired and fixed networks, for which TCP has been originally designed. In these networks, congestion is the main cause of packets losses and the overall path delay does not change too fast. Congestion control mechanisms in the standard TCP, based on these assumptions, work poorly on networks that include terrestrial/satellite wireless hops. The characteristics of these networks that impact heavily on end-to-end TCP performance are:

- High error rate (e.g., due to fading, interference, etc.);
- High latency (e.g., in long RTT satellite links);
- Variable delays and data rate (due to varying channel conditions);
- Intra- and intersystems handovers (due to mobility).

High error rates introduce another cause of packet loss. The TCP sender detects a packet loss in two different ways:

- Retransmission timeout;
- Duplicate ACKs.

TCP sender sets a timeout when it sends data and assumes that the packet is lost, either if the acknowledgment has not been received when the timeout expires or if three duplicate ACKs have been received. Since TCP cannot distinguish between congestion-related and non-congestion-related packet loss, the TCP sender interprets a packet loss as congestion signal and enters into a congestion avoidance state that can substantially reduce the overall throughput. Due to the time-varying channel conditions (fading and interference level can dynamically change) and to mobility, the sender can experience a sudden change in the available bandwidth. For example, a connection can change cell and experience a sudden increase of the available bandwidth, which will be underutilized because the linear probing algorithm of the congestion avoidance phase only slowly increases the transmission rate. Furthermore, in the new cell the RTT could be longer and, before TCP changes its estimate of the RTT, in order to properly set RTOs, some spurious timeouts can occur. The handover procedure itself can result in a sudden increase of the path latency. During a handover procedure, the mobile terminal in required to perform some time-consuming actions before packets can be transmitted in a new cell. In some cases, the link experiences delays that exceed the typical RTT by several times, thus causing spurious timeouts and unnecessary retransmissions. In particular, intersystems handover can adversely affect TCP connections, since many features that are negotiated at the connection establishment and cannot be changed later could not be optimized for the new link characteristics.

4.3 TCP Enhancements

Improving TCP throughput over high-speed networks and terrestrial/satellite wireless links continues to be an active area of research and discussion [6–8]. Most of the work developed in the frame of satellite networks, which fall in the LFNs regime, is a good starting point also for LTNs, such as 3G cellular networks.

This section discusses the main configuration options recommended in more modern versions of TCP. All solutions have been derived moving from the awareness that a dramatic change of TCP just to support mobile users and wireless links is not practical. In fact, it would imply major changes to the whole structure of routers and hosts that have been already installed. Therefore, every TCP enhancement has to remain compatible to the standard TCP in fixed networks.

In Table 4.1, a list of enhancements is shown, highlighting their main features.

4.3.1 Summary of Recommendations

In this section, a set of recommendations for supporting TCP connections over high-speed networks and terrestrial/satellite wireless links is presented. These configuration options are all found in modern TCP stacks. Two examples of applications that adopt TCP as their transport protocol and recommend TCP optimization mechanisms closely aligned with the existing recommendations are the Wireless Applications Protocol (WAP) technology and the i-mode service in Japan.

Table 4.1
Enhancements to Standard TCP

Extensions	Comments
Fast retransmit/fast recovery	It allows a quick recovery from noncongestion-related losses without giving up the safety provided by the congestion control mechanism
Explicit loss notification (ELN)	Efficient recovery from link outages due to handovers
Large window size	It is necessary to achieve high data rates. Window size limitations: receiver buffer size
Large initial window	It reduces the slow start transient. It is effective for transfer of small amounts of data
Selective acknowledgments (SACKs)	It enables a sender to continue to transmit segments without entering into a time-consuming slow-start phase. It is more useful in LFNs, especially if large windows are being used.
Splitting TCP	+ Each connection can be optimized independently
	+ Transmission errors on the satellite links cannot propagate into the fixed network
	− Loss of the original end-to-end semantic
	− Extra overhead in moving TCP states in case of handoff
Snooping TCP	+ Preservation of the original end-to-end semantic
	+ No extra overhead in moving TCP states in case of handoff
	− No good isolation of the wireless link

4.3.1.1 Fast Retransmit/Fast Recovery

Fast retransmit/fast recovery mechanisms enable TCP sender to recover rapidly from a single lost packet, or one that is delivered out-of-sequence, without shutting down the CWND. The *fast retrasmit* mechanism is based on the notion that when a sender receives continuous acknowledgments for the same packet, it can infer that the packet loss was not congestion-related. In fact, receiving acknowledgments shows that the receiver is continuously receiving from the sender. Therefore, when the TCP sender receives three duplicated ACKs, it retransmits the lost packet before the timer expires, without waiting for the full timeout, thus saving time. Furthermore, since the reception of subsequent segments, which are generating the duplicated ACKs, was successful, there is no need to enter the slow start.

Fast recovery allows the sender to transmit at half its previous rate (regulating the growth of its window, based on congestion avoidance), rather than beginning a slow start. When using these mechanisms, TCP only interprets a timeout due to a missing acknowledgment as network congestion. The fast retransmit algorithm first appeared in the 4.3BSD TCP Tahoe release, and it was followed by slow start. The fast recovery algorithm appeared in the 4.3BSD TCP Reno release [9].

4.3.1.2   Explicit Loss Notification

Frequent handovers could cause link outage (i.e., a long period where all frames are lost). If the outage lasts for a longer period than the TCP RTO, congestion control mechanisms will trigger transport layer retransmissions and will reduce its CWND to one segment. Furthermore, TCP only determines that the outage is over upon reception of an ACK. In order to improve the efficiency during handover procedures, the sender should start transmitting again as soon as the outage is over, without reducing the CWND. With this aim, explicit loss notification (ELN) and fast retransmission mechanisms are exploited [10–12]. In this scheme, the network layer informs the TCP receiver that the handover procedure is over. Then, the receiver sends to the sender three duplicate NACKs, to indicate the end of the handover procedure, containing information about the last received packets. The sender starts the handover recovery phase: it sends three successive packets to probe the link and adjust the transmission window before retransmitting lost packets.

4.3.1.3   Large Window Size

With a small window size, it is impossible for TCP to achieve a high data rate. Many implementations do not offer a very large window size (few kilobytes is typical). A first limitation to the window size is represented by the buffer limitations at the receiver. Even without this limitation, however, the standard TCP window size cannot exceed 64 kB, because the field in the TCP header devoted to window advertisement is only 16 bits wide. This limits the effective TCP bandwidth to 64 kB divided by the RTT. In case of a 0.5s RTT GEO link, the maximum achievable throughput is 1 Mbps. In other words, without a window larger than 64 kB, a TCP sender can only transmit 1 Mbps of data before waiting for an acknowledgment from the sender. RFC 1323 defines a set of window scaling options that are available for LFNs [13]. For LFNs, windows large enough are required to utilize the available network bandwidth. The so-called long thin networks, such as 2.5G networks characterized by a smaller BD product around 1 to 5 kB, need larger windows in order to avoid retransmission timeouts in the presence of packet losses [4].

4.3.1.4   Large Initial Window

In order to reduce the slow start transient, there is the option of allowing TCP sender to transmit more than one segment at the beginning of the initial slow start. If there is capacity in the path, such an option reduces the slow start by up to three RTT. The increased initial window mechanism is effective for applications that usually involve transfers of small amounts such as typical Internet applications enabled for mobile wireless devices. An initial window of two segments is recommended in RFC 2581 [14]. RFC 2414 also considers the use of an initial window size larger than two segments [15].

4.3.1.5 Selective Acknowledgments

Standard TCP is only able to acknowledge data received in order. The TCP sender can only learn about a single lost packet per RTT. Selective acknowledgments (SACKs) enable TCP to acknowledge data received out of order [16]. The sender can then retransmit only the missing data segments thus reducing the retransmission period. The SACK option is effective when multiple TCP segments are lost in a single window. In this case TCP SACK enables a sender to continue segments transmission without entering into a time-consuming slow start phase. SACK is more useful in LFNs, especially if large windows are being used, because there is a considerable probability of multiple segment losses per window. This is not the case in many LTNs [4], where TCP windows are much smaller, and burst errors due to fading should be much longer to damage multiple segments. Therefore, SACK with large TCP windows is recommended in satellite environments [17], whereas in LTNs this extension should be supported mainly for compatibility with TCP extensions for non-LTNs, although in many cases they do not benefit from it. Moreover, it has been shown that SACK improves throughput for *Snooping TCP* enhancement (see Section 4.3.3) when multiple segments are lost per window [18].

The TCP extensions described above can solve some of the limitations of standard TCP, but other problems that arises in wireless links, such as those related to the higher error rate or the mobility, are not addressed effectively.

4.3.2 Splitting TCP

Splitting TCP, also called *cascading TCP* X*O*A*, divides a TCP connection into a fixed part and a wireless part. The idea behind this solution is that TCP running over the wireless link can be modified and optimized to the link characteristics. Figure 4.3 shows an example where the wireless link is a satellite link.

Standard TCP is used between the fixed computer and the gateway (GTW). The connection is terminated at the GTW. Over the wireless link, a proprietary transport protocol, adapted to the link, can be used, although it is not a requirement. In [19], a Satellite Transport Protocol (STP) is designed to be used in the satellite hop of a TCP splitting approach. Splitting the TCP connection can give benefits even without changing the TCP over the wireless link. In fact, in this configuration, transmission errors over the wireless link cannot

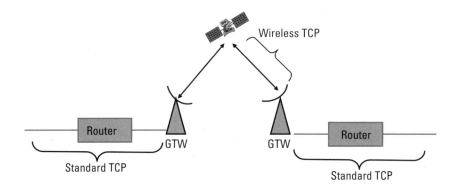

Figure 4.3 Splitting TCP approach in a network including satellites.

propagate into the fixed network. If the fixed computer sends a packet, the access point (GTW in the example) acknowledges this packet and then tries to forward it through the wireless link. If the packet is lost in the wireless link, the access point tries to retransmit the packet locally and the fixed host would not notice the packet loss. Furthermore, these local retransmissions experience a shorter delay, thus allowing standard TCP to recover faster from packet losses. The main drawbacks of splitting TCP are as follows:

- *The loss of TCP end-to-end semantics.* A sender, when receiving an acknowledgment, assumes that the packet has been received success-fully by the destination host, while this only means that the access point has received it successfully. Therefore, crashes of the intermediate nodes become unrecoverable. Moreover, the end-to-end usage of IP layer security mechanisms (see Chapter 3) is disabled. If users apply end-to-end encryption [20], the point partitioning the TCP connection has to be integrated into all security mechanisms.

- *Increased handover latency, which could trigger unnecessary retransmissions.* Let us assume that the connection between the mobile host and the correspondent fixed host is segmented at the foreign agent of mobile IP (see Chapter 3). This is an effective solution, since the foreign agent controls mobility of the mobile host anyway. The access point partitioning the TCP connection has to buffer all packets sent by the correspondent fixed host before forwarding them to the mobile host. If the mobile host performs a handover to another foreign agent, the old foreign agent has to forward the buffered data to the new one before it can start forwarding them to the mobile host. This process might take a long time and new packets may arrive in the meantime.

4.3.3   Snooping TCP

Snooping TCP, described in [21], consists of placing a router close to the mobile host, which buffers data and performs fast local retransmissions of packet loss, thus giving the illusion to the sender of a shorter path delay. Again, this TCP enhancement could be implemented properly in the foreign agent of Mobile IP. The connection in this case is not segmented (i.e., it is not terminated at the foreign agent and the end-to-end TCP semantics is not violated) (see Figure 4.4).

In fact, the foreign agent does not acknowledge data to the correspondent node, but it just "snoops" the packet flow in both directions to recognize acknowledgments [22, 23]. Let us consider first the transmission direction from the correspondent node to the mobile host. The foreign agent buffers all the packets transmitted by the correspondent node until it receives the ACK from the mobile host. If the foreign agent does not receive an ACK from the mobile host before a certain amount of time, it retransmits the packet directly from the buffer, thus performing faster retransmissions than from the correspondent node. Furthermore, the foreign agent suppresses the duplicate acknowledgments on their way from the receiver back to the sender, in order to avoid unnecessary retransmissions of data from the correspondent host. The timeouts for acknowledgments at the foreign agent can be set much shorter, since they depend only on the sum of the single-hop propagation delay over the wireless link and the processing delay. On the other hand, the timeouts of the correspondent node still work and can trigger a retransmission even if the foreign agent crashes. Duplicate packets, already retransmitted locally and acknowledged by the mobile host, are discarded to avoid unnecessary traffic on the wireless hop. In the opposite direction, the foreign agent sends back to the mobile host a NACK as soon as it detects a missing packet in the received stream.

Snooping TCP represents a simple solution with the following main advantages:

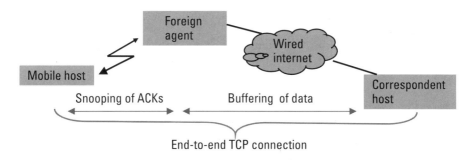

Figure 4.4 Snooping TCP approach.

- It preserves the end-to-end semantics (i.e., crashes of intermediate nodes are recoverable).

- It requires no changes at the correspondent host (fixed computer).

- If the foreign agent does not use the enhancement (for instance, the new foreign agent after a handover) or it crashes, the approach falls back automatically to the standard TCP.

- The extra time needed, during the handover procedure in the splitting TCP, to transfer all the buffered packets to the new foreign agent is not needed anymore. If some data are not transferred to the next foreign agent, the timeout at the correspondent node will trigger a retransmission of these packets, possibly already at the new care-of address.

This approach, however, does not provide a complete isolation from the wireless link. If the foreign agent takes too much time to transmit successfully the packet over the wireless link, the timeouts at the correspondent node will expire, thus triggering a retransmission and activating the congestion avoidance mechanism. Furthermore, it is still not clear how this scheme can be used together with encryption end-to-end mechanisms. If the TCP header is encrypted, snooping on the sequence number to detect a loss will not work any longer [20].

4.4 Data Link Layer Approach

Transport layer approaches, such as snooping TCP or split connection, basically try to make the sender aware that some packet losses are not congestion-related. Data link layer approaches, which are presented in this section, try transparently to enhance the performance by making the TCP layer "see" a more reliable channel.

The link layer mechanisms that have the largest effect on TCP performance over wireless links are error recovery schemes such as data link layer ARQ and FEC coding [24].

FEC coding is commonly included in the physical layer design of wireless links and may be used simultaneously with link ARQ. In FEC, redundancy is added at the transmitter and is used at the receiver to correctly recover the information, even in the presence of some transmission errors. In ARQ, a smaller percentage of redundancy than in FEC is added to the data, which makes it possible only to *detect* errors. A return channel (Rx = Tx) is necessary to feedback the status of the received information. Detected lost frames are then retransmitted, until they are successfully received or they are discarded because the allowed overall transmission delay for packet transmission has been reached. Recent studies suggest that Hybrid ARQ [25–31] schemes, which combine adaptive

FEC with link ARQ procedures to reduce the probability of retransmitted packet loss, might be more appropriate than pure FEC or ARQ schemes for a wireless network that carries traffic with different characteristics and QoS requirements. Adaptive FEC schemes that combine modulation and coding with hybrid ARQ schemes are presented in more detail in Chapter 5.

In what follows, basic ARQ schemes are summarized and their interactions with transport protocols performance are highlighted, in order to provide hints on the choice of protocol parameters. Further comments on the data link layer protocols design are given in Section 4.5.

4.4.1   Data Link Layer ARQ Protocols

The packet is divided into link layer frames that may contain all, or part of, one or more IP packets. Cyclic redundancy check (CRC) bits are added to each frame and are used at the receiver to detect whether a frame has been received with any error. The frame loss detection at the sender, which has to schedule the frame retransmission in case of errors, may be via a link protocol timer, by detecting positive link acknowledgment frames, by receiving explicit negative acknowledgment frames and/or by polling the link receiver status.

Regardless of the mechanism applied for loss detection, there are two main categories of ARQ schemes that are widely used:

- Stop-and-wait;
- Sliding-window.

Stop-and-wait schemes are simple but inefficient. The sender transmits a single frame and waits for an acknowledgment of this frame. Upon reception of the ACK, it transmits the next frame or retransmits the same frame, in case of loss. The sliding-window mechanism has already been described in this chapter (Section 4.2.1), applied at transport layer. In an ARQ sliding-window protocol, each frame is assigned a unique sequence number. A transmission window size of N frames means that N unacknowledged frames can be sent. Among the sliding-window ARQ schemes two techniques are widely used:

- Go-back-N;
- Selective repeat.

4.4.1.1   Go-back-N

The transmitter starts sending W frames, with sequence number 0, 1,..., $W-1$, and sets a timer. W is the transmission window at link layer. As soon as the sender receives the acknowledgment of the reception of the frame n, it transmits the frame with sequence number ($W + n$). In the basic version of the protocol, if the ACK of frame n has not been received when the timer reaches a given retransmission timeout, T_{RTO}, the frame n is immediately retransmitted together with the frame ($n + 1$), ..., ($n + W$). T_{RTO} should be set such that: $T_{RTO} \geq W^*t$, where t is the frame transmission delay. Figure 4.5 shows the operation of a go-back-N (GBN) algorithm.

In this figure, three situations are shown:

(a) Frame 0 has not been received and no ACK has been sent.

(b) When timeout T_{RTO} expires, frame 0 is retransmitted together with all the successive ($W - 1$) frames ($W = 3$).

(c) All frames are received successfully.

The receiver has to be able to discard multiple copies of the same frames, received correctly after the first incorrect frame. This scheme is more efficient than the stop-and-wait one, since more than one frame can be sent before receiving an ACK. A more efficient version of GBN uses NACK to inform the sender that a frame has not been received. In such a way, the sender may retransmit the lost frame [and all the ($W - 1$) successive frames] as soon as it gets the NACK without waiting for the timeout to expire. Retransmission timeouts should be set in any case, to protect the protocol with respect to the loss of NACKs.

4.4.1.2   Selective Repeat

The selective repeat (SR) scheme avoids the unnecessary retransmissions of the GBN scheme. As in the GBN, the transmitter sends all frames in one transmission window W and waits for acknowledgments. If the ACKs with sequence

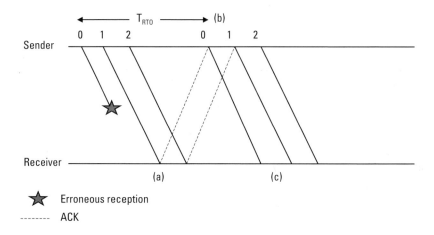

Figure 4.5 GBN ARQ scheme.

number 0 to L are received, the sender is allowed to transmit frames from $L + 1$ to $L + W$. If the ACK of the frame n is not received before the T_{RTO} expires, only the lost frame is retransmitted, as is shown in Figure 4.6.

Let us denote with H the sequence number of the last frame received correctly. If the receiver receives a correct frame with sequence number between [$H + 1$; $H + W$], then it has to buffer the frame if it is out-of-sequence, or it can pass it to the upper layer if it is in-sequence. Otherwise, if the sequence number of the last received frame does not fall in that range, the receiver has to discard the frame and send an ACK. Therefore, the receiver has to be able to buffer $W - 1$ frames arriving out of order. Also, the SR scheme can be modified by introducing NACKs.

Four situations are shown in Figure 4.6:

(a) Frames 0 and 2 are not correctly received and no ACKs are sent.

(b) Lost packets are retransmitted together with other packets ($W = 4$).

(c) The first five frames are received successfully.

(d) Upon reception of the ACKs for the five frames, successive frames are sent.

SR-ARQ is much more complex than GBN also because packet reordering should be done at the data link layer. In fact, in the SR-ARQ scheme, packets may arrive out of order and if reordering is not performed at the data link layer, they will arrive out of order at the TCP layer thus triggering the TCP fast retransmit procedure.

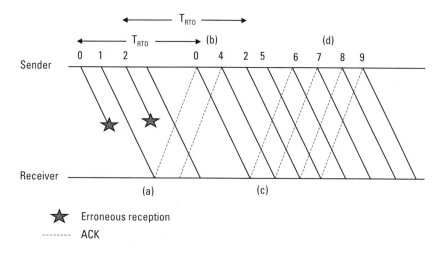

Figure 4.6 Selective repeat ARQ scheme.

4.4.1.3 ARQ Protocols Design Parameters

Local retransmissions across a single link have a lower delay than the end-to-end TCP acknowledgment control loop. Therefore, ARQ is able to provide faster retransmissions of lost TCP segments, at least for a reasonable number of retries. Excessive retransmissions could result in:

1. TCP retransmissions timeouts (RTO) expiring;
2. The selection of a larger TCP RTO due to the higher overall path delay experienced at TCP layer.

Both those effects will result in the degradation of the overall throughput performance. Furthermore, applications performance is determined by the cumulative delay of the entire end-to-end Internet path. This path may include an arbitrary or even a widely fluctuating number of links, where any link may or may not use ARQ. The delay introduced by ARQ protocols in each link can be highly variable, depending on link conditions. As a result, it is not possible to provide clear upper bounds on the acceptable delay that a link can add to a path, which is very important in case of delay-sensitive applications. Therefore, it is necessary to control the maximum delay introduced in the frame transmission by the ARQ protocol. An important parameter of ARQ protocols is the *persistency*, which defines how long the link is allowed to delay the frame before giving up and discarding it. The *persistency* can be measured in milliseconds or in maximum number of retransmissions. A low persistency (2–5 retransmission attempts) reduces the effectiveness of ARQ protocols in terms of reliability, but provides a lower degree of delay jitter and reduces the interaction with the TCP RTO timers. As a consequence, the probability that duplicate copies of the same frames be transmitted decreases. High-persistent ARQ schemes may benefit a TCP session in presence of links that experience persistent loss, where many consecutive frames are corrupted over an extended time (due to fading, handover). During this loss event, there is an increased probability that a retransmitted frame is also lost. In this case, the link ARQ persistency should be longer than the largest link loss event. Examples of low persistence ARQ protocols are found in satellite systems [32–34]. Actually in such long RTT links, FEC coding, which does not interact with retransmission mechanisms at TCP layer, has been widely recommended [35]. In the RLC protocol used in the W-CDMA standard, which is a selective repeat sliding window ARQ [36], the maximum number of retransmissions is a reconfigurable parameter specified by the RRC (see Chapter 1) through RLC connection initialization [37]. RRC can set the maximum number of retransmissions up to 40. Therefore, RLC can be configured as high- or low-persistence ARQ protocol (never as a perfect-persistence ARQ protocol).

Another important ARQ parameter that impacts TCP throughput performance is *frame size*. A frame size much smaller than IP packet size introduces more frame header overhead per packet. On the other hand, using smaller frames may improve the efficiency of the ARQ process as it increases the probability of retransmitting only the erroneous parts of the IP packet. This efficiency improvement decreases if the channel is characterized by long burst of errors where one or more packets may be corrupted. Therefore, the optimum frame size is strictly dependent on the channel burstiness. Note that in radio channels, interleaving is typically used to reduce the burstiness of the errors. However, the introduction of the interleaver increases the processing time at the receiver as a packet cannot be decoded until a certain number of other packets, which depends on the interleaver depth, has been received. With high data rate and low mobility users, the interleaver depth that is needed to assume channel errors completely uncorrelated could be too high. Therefore, more realistic performance assessments should consider imperfect interleaving and, hence, residual channel error correlation.

4.5 New Trends in the Wireless Networks Design

Data link and transport layer protocols work on a number of successive data blocks, and, hence, the simple specification of the average error rate is no longer sufficient to assess protocol performance. The common layered design approach relies on the assumption that, whatever scheme or protocol is going to be implemented at data and transport layer, lowering the average error rate at physical layer will always result in better performance for the upper layer protocols. Therefore, the main task of the physical layer designer is to lower the average error rate; the data link layer designer focuses on the reliable transfer of frames; routing and mobility management are the most important issues for the network layer designers, while transport layer designers focus on the end-to-end performance. This approach is no longer valid in a wireless scenario with highly time-varying channels. It is widely recognized that link characteristics, such as channel characteristics as well as physical and data link layer mechanisms, have to be carefully considered in the assessment of end-to-end performance [38–43]. The first step of the assessment is a proper characterization of the wireless channel at packet level. The average error rate is an incomplete performance metric. Several studies have shown the strong dependence of error statistics, especially of second-order statistics, on data link and transport layer protocols performance [44, 45]. Two consequences of neglecting the channel autocorrelation are:

1. The inability to predict the protocol performance, coming up with wrong conclusions in some cases;

2. The waste of the opportunity to optimize data and transport layer protocols design, according to the knowledge of the channel.

Throughput performance of TCP over bursty channel can be found in [45–50]. Figure 4.7 shows the dependence of throughput performance on the channel burstiness.

The figure shows the TCP throughput performance as a function of the average frame error rate (FER), ε, for i.i.d errors and for burst errors with different values of the average burst length. The figure highlights that for smaller FER, where the protocol performance is dominated by the buffer size as packet losses occur mostly due to congestion and channel statistics is almost irrelevant; whereas for higher FER, performance is dominated by the error statistics and depends strongly on error correlation. Furthermore, a higher degree of burstiness results in a TCP throughput performance improvement. In Figure 4.7, the Tahoe version of TCP is considered. In [48], it is shown that the burstiness of the channel improves the performance of TCP NewReno on slow fading channel, provided that the receiver buffer size be in excess of the average packet error burst length. Moreover, in these highly correlated channels, TCP NewReno does not outperform TCP Tahoe, basically because all improvements have been designed taking into account a lossy channel with random errors and not bursty errors. These results lead to a fundamental conclusion: from the data link-transport protocol point of view, a proper design of the physical layer should

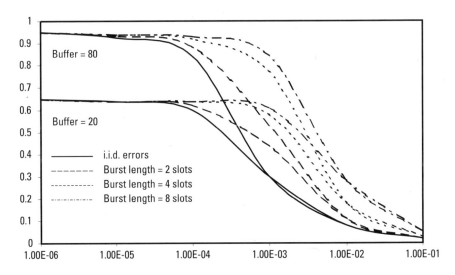

Figure 4.7 Lower bounds for the throughput performance of TCP Tahoe, η, versus the average packet error rate on the channel, ε, for independent and identically distributed (i.i.d) errors and correlated errors with average burst length 2, 4, and 10 slots. Buffer size B = 20 and 80, RTT = 100 slots. (*From:* [44].)

exploit the bursty characteristics of the channel instead of destroying it (for example, by using interleavers).

A more general conclusion is that a good end-to-end wireless network performance is not possible without a truly optimized, integrated, and adaptive network design, where each level in the protocol stack should adapt to wireless link variations in an appropriate manner, taking into account the adaptive strategies at the other layers.

In what follows, some of the design protocols proposals that try to exploit the knowledge of the error statistics are presented.

4.5.1 Improved Link Layer Schemes

The ARQ protocol proposed in [51] uses a type of time diversity. The receiver waits for a period of time t before retransmitting an unacknowledged frame. In fading channel, choosing t based on the knowledge of the feedback channel burstiness [52] may avoid retransmissions of large data blocks in case of deep fades. An improvement can be achieved by tuning the protocol parameters according to information about link conditions that are available at the receiver. In [53], a retransmission scheme has been proposed where the receiver directly informs the sender about the last packets received in sequence, the last packet received, and whether or not the other packets in between were received, by periodically transmitting packets to the sender. The scheme can improve efficiency by using this information to retransmit preemptively some unacknowledged frames. Of course, the feedback packets use a portion of the reverse link capacity.

An alternative approach consists of using some indirect measurement of channel conditions, such as the number of ACKs and NACKs received within a certain optimized observation period. In [54], a stop-and-wait protocol, that updates packet lengths according to this indirect information on the channel conditions is proposed. Other approaches try to increase the reliability of retransmissions, accounting that if one frame has been lost the channel is bad. This can be achieved by adding more redundancy to the retransmitted frame. In [55], a power control scheme for a CDMA system is proposed where the transmitter raises its output power for a retransmitted radio link frame. The provided higher reliability of the retransmissions reduces the average time the protocol takes to deliver a packet through the radio link layer, at the price of raising the average transmit power per user. Closed power control (see Chapter 3) is a feedback already included at physical layer of CDMA system, which can provide the sender with information about the channel conditions. In [56], it is shown how a truncated power control can be used to introduce some *time diversity*. The latter allows for an increase in the reliability of the channel at link layer without increasing the average transmission power, thus increasing the effectiveness of ARQ protocols in improving end-to-end TCP performance.

In fact, instead of transmitting and then retransmitting, it is better, in terms of delay, to avoid transmitting and to wait until the channel condition changes. Bad channel conditions are assumed when a high (higher than a proper chosen power threshold) transmission power level is required by the sender in order to compensate for the channel and the interference level. Therefore, the proposed strategy consists of combining the following steps:

- Decrease the FER at the physical layer by increasing the target signal-to-noise ratio, SNR_{req};
- Adopt the following truncated power management policy:
- Transmit if required transmission power is lower than a power threshold;
- Idle otherwise.

Figure 4.8 shows the TCP/Radio Link Layer (RLL) protocol stack and the effect of the truncated power control in the frame transmission.

The two feedback mechanisms at RLL and physical layer are here highlighted: the ARQ mechanism and the closed loop power control, respectively. If a power threshold is introduced, it is expected to idle during deep fade or high interference levels, waiting to transmit when the channel attenuation and/or the

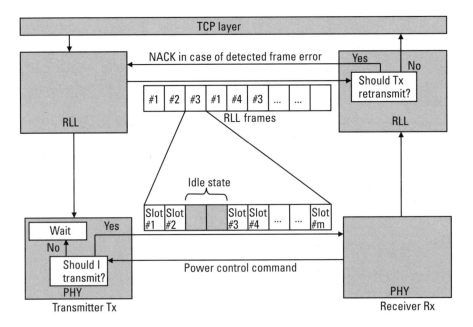

Figure 4.8 TCP/RLL protocol stack with a truncated power control.

interference level is lower. Thereby, a delay at physical layer is introduced. The key point is that in spite of that delay increase at the physical layer, the overall delay at RLL, depending also on the reliability of the physical channel, could be lower. The strategy turns out to be very effective when the channel is not correlated since it introduces a type of time diversity. It is also more effective in presence of a longer frame size and low interference load. In any case, it provides a way to trade-off the transmit powerand the delay, hence offering a further element of flexibility in the system design.

Other innovative ARQ schemes are described in [57–59].

4.6 Conclusions

State-of-the-art enhancements proposed and implemented for improving TCP performance over wireless links have been presented in this chapter. One of the most promising areas that can be identified is the research of innovative ARQ schemes able to exploit the channel knowledge. It is also important to be aware that link ARQ is just one method of error recovery, and that other complementary physical-layer techniques may be used instead of, or together with, ARQ to improve overall link throughput for IP traffic. The choice of potential schemes includes adaptive modulation, power control schemes, adaptive FEC, and interleaving. Several studies have been already undertaken in this direction, as proved by the work of IETF on the implications of link characteristics on the end-to-end performance. As stated in this work, there is a need for more research to more clearly identify the importance of the interdependencies between the physical layer and higher layers over various types of channels. The importance of taking into account the interaction of the physical layer and higher layers, in particular the MAC layer, has been already outlined in Chapter 2. Researchers and implementers should clearly indicate the lossy channel model, link and end-to-end path delays, the performance metrics to be measured at each layer of the protocol stack, and the adaptive strategies to be developed for each layer to better respond to the variations of these performance metrics in a new fully integrated design approach.

References

[1] Stevens, W. R., *TCP/IP Illustrated, Volume 1: The Protocols*, Reading, MA: Addison Wesley Longman Publishing Company, 1994.

[2] Khafizov, F., and M. Yavuz, "Running TCP over IS-2000," in *Proc. of IEEE ICC 2002*, pp. 3444–3448.

[3] TIA/EIA/IS-707-A-2.10, "Data Service Options for Spread Spectrum Systems: Radio Link Protocol Type 3," Jan. 2000.

[4]   Montenegro, G., et al., "Long Thin Networks," RFC 2757, Jan. 2000.

[5]   Allman, M., et al., "Ongoing TCP Research Related to Satellites," RFC 2760, Feb. 2000.

[6]   Patridge, C., and T. J. Shepard, "TCP/IP Performance over Satellite Links," *IEEE Network*, Sept.–Oct. 1997, pp. 44–49.

[7]   Metz, C., "TCP over Satellite—The Final Frontier," *IEEE Internet Computing*, Jan.–Feb. 1999, pp. 76–80.

[8]   Lakshman, T. V., and U. Madhow, "The Performance of TCP/IP for Networks with High Bandwidth Delay Products and Random Loss," *IEEE Transactions on Networking*, Vol. 5, No. 3, June 1997, pp. 336–350.

[9]   Floyd, S., and T. Henderson, "The NewReno Modification to TCP's Fast Recovery Algorithm," RFC 2582, April 1999.

[10] Chan, A., D. Tsang, and S. Gupta, "Impacts of Handoff on TCP Performance in Mobile Wireless Computing," *IEEE International Conference in Personal Wireless Communication*, 1997, pp. 184–188.

[11] Manzoni, P., D. Ghosal, and G. Serazzi, "Impact of Mobility on TCP/IP: An Integrated Performance Study," *IEEE Journal on Selected Areas on Communication*, Vol. 13, No. 5, June 1995.

[12] Caceres, R., and L. Iftode, "Improving the Performance of Reliable Transport Protocols in Mobile Computing Environments," *IEEE Journal on Selected Areas on Communication*, Vol. 13, No. 5, June 1995.

[13] Jacobson, V., R. Branden, and D. Borman, "TCP Extensions for High Performance," RFC 1323, IETF Network Working Group, May 1992.

[14] Allman, M., V. Paxson, and W. Stevens, "TCP Congestion Control," RFC 2581, April 1999.

[15] Allman, M., S. Floyd, and C. Patridge, "Increased TCP's Initial Window," RFC 2414, Sept. 1998.

[16] Mathis, M., and J. Madhavi, "Forward Acknowledgment: Refining TCP Congestion Control," *Proc. ACM SIGCOMM'96*, Aug. 1996, pp. 281–291.

[17] Allman, M., C. Hayes, and S. Ostermann, "An Evaluation of TCP with Larger Initial Windows," *Computer Communication Review*, Vol. 28, No. 3, July 1998.

[18] Balakrishnan, H., et al., "A Comparison of Mechanisms for Improving TCP Performance over Wireless Links," *IEEE/ACM Transactions on Networking*, Vol. 5, 1997, pp. 756–769.

[19] Henderson, T. R., and R. H. Hatz, "Transport Protocol for Internet Compatible Satellite Networks," *IEEE Journal on Selected Areas on Communication*, Vol. 72, No. 2, Feb. 1999, pp. 326–344.

[20] Atkinson, R., "IP Encapsulating Security Payload," RFC 1827, 1995.

[21] Liu, Z., et al., "Evaluation of TCP Vegas: Emulation and Experiment," *Proc. ACM SIGCOMM '95*, Aug. 1995, pp. 185–196.

[22] Balakrishnan, H., S. Seshan, and R. H. Katz, "Improving Reliable Transport and Handoff Performance in Cellular Wireless Networks," *Wireless Networks, J. C. Baltzer*, Vol. 1, 1995.

[23] Brewer, E. A., et al., "A Network Architecture for Heterogeneous Mobile Computing," *IEEE Personal Communications*, Vol. 5, 1998.

[24] Lin, S., and D. Costello, *Error Control Coding: Fundamentalsadn Applications*, Englewood Cliffs, NJ: Prentice-Hall, 1983.

[25] Liu, H., El-Zarki, "Performance of H.263 Video Transmission over Wireless Channels Using Hybrid ARQ," *IEEE Journal on Selected Areas on Communication*, Vol. 15, Dec. 1997, pp. 1775–1786.

[26] Rice, M., and S. B. Wicker, "Adaptive Error Control for Slowly Varying Channels," *IEEE Transanction on Communication*, Vol. 42, Feb./Mar./Apr. 1994, pp. 917–925.

[27] Costello, D., et al., "Applications of Error-Control Coding," *IEEE Trans. Inform. Theory*, Vol. 44, Oct. 1998, pp. 2531–2560.

[28] Link, R., S. Kallel, and S. Bakhtiyari, "Throughput Performance of Memory ARQ Scheme," *IEEE Trans. Veh. Tech.*, Vol. 48, May 1999, pp. 891–899.

[29] Krunz, M. M., and J. G. Kim, "Fluid Analysis of Delay and Packet Discard Performance for QoS Support in Wireless Networks," *IEEE Journal on Selected Areas on Communication*, Vol. 19, No. 2, Feb. 2001, pp. 384–395.

[30] Caire, G., and D. Tninetti, "The Throughput of Hybrid-ARQ Protocols for Gaussian Collision Channel," *IEEE Trans. Inform. Theory*, Vol. 47, No. 5, July 2001, pp. 1971–1988.

[31] Prasad, A. R., Y. Shinohara, and K. Seki, "Performance of Hybrid ARQ for IP Packet Transmission on Fading Channel," *IEEE Trans. Veh. Tech.*, Vol. 48, No. 3, May 1999, pp. 900–910.

[32] Cheng, H. S., et al., "An Efficient Partial Retransmission ARQ Strategy with Error Codes by Feedback Channel," *IEE Proceedings*, Vol. 145, No. 5, 2000, pp. 263–268, 2000.

[33] Ward, C., et al., "A Data Link Control Protocol for LEO Satellite Networks Providing a Reliable Datagram Service," *IEEE/ACM Trans. on Networking*, Vol. 3, No. 1, Feb. 1995, pp. 91–103.

[34] Samaraweera, N., and G. Fairhurst, "Robust Data Link Protocols for Connection-Less Service over Satellite Links," *Int. Journ. Sat. Commun.*, pp. 427–437, 1996.

[35] Allman, D., D. Glover, and L. Sanchez, "Enhancing TCP over Satellite Channels Using Standard Mechanisms," RFC 2488, Jan. 1999.

[36] Third Generation Partnership Project, "RLC Protocol Specification, (3G TS 25.322)," 1999.

[37] Third Generation Partnership Project, "RRC Protocol Specification, (3GPP TS 25.331)," Sept. 2001.

[38] Fairhurst, G., and L. Wood, "Link ARQ Issues for IP Traffic," Internet draft, Nov. 2000, http://www.ietf.org/internet-drafts/draft-ietf-pilc-link-arq-issues-03.txt, work in progress.

[39]   Dawkins, S., and G. Montenegro, "End-to-End Performance Implications of Slow Links," RFC 3150/BCP 48, July 2001.

[40]   Karn, P., et al., "Advice for Internet Subnetwork Designers," Internet draft, Nov. 2000, http://www.ietf.org/internet-drafts/draft-ietf-pilc-link-design-06.txt, work in progress.

[41]   Dawkins, S., et al., "End-to-end Performance Implications of Links with Errors," RFC 3135/BCP 50, Aug. 2001.

[42]   Gurtov, A., "Making TCP Robust Against Delay Spikes," University of Helsinki, Department of Computer Science, Series of Publications C, C-2001-53, Nov. 2001, http://www.cs.helsinki.fi/u/gurtov/papers/report01.html.

[43]   Balakrishnan, H., et al., "TCP Performance Implications of Network Asymmetry," Internet draft, Sept. 2001, http://www.ietf.org/internet-drafts/draft-ietf-pilc-asym-07.txt.

[44]   Zorzi, M., and R. R. Rao, "Perspective on the Impact of Error Statistics on Protocols for Wireless Networks," *IEEE Personal Communications*, Oct. 1999, pp. 32–40.

[45]   Zorzi, M., R. R. Rao, "The Effect of Correlated Errors on the Performance of TCP," *IEEE Comm. Letters*, Vol. 1, No. 5, Sept. 1997, pp. 127–129.

[46]   Zorzi, M., R. Rao, and L. B. Milstein, "Error Statistics in Data Transmission over Fading Channels," *IEEE Trans. on Comm.*, Vol. 46, No. 11, Nov. 1998, pp. 1468–1477.

[47]   Zorzi, M., R. R. Rao, "On the Statistics of Block Errors in Bursty Channels," *IEEE Trans. on Comm.*, Vol. 45, No. 6, June 1997, pp. 660–667.

[48]   Chockalingam, A., M. Zorzi, and R. R. Rao, "Performance of TCP on Wireless Fading Links with Memory," *Proc. ICC '98*, June 1998.

[49]   Zorzi, M., A. Chockalingam, and R. R. Rao, "Performance Analysis of TCP on Channels with Memory," *IEEE Journal on Selected Areas on Communication*, Vol. 18, No. 7, July 2000, pp. 1289 –1300.

[50]   Chuan, M. C., O.-C. Yue, and A. DeSimone, "Performance of Two TCP Implementations in Mobile Computing Environments," *Proc. IEEE Globecom*, Singapore, Nov. 1995.

[51]   Kim, S. R., and C. K. Un, "Throughput Analysis for Two ARQ Schemes Using Combined Transition Matrix," *IEEE Trans. on Comm.*, Vol. 40, Nov. 1992, pp. 1679–1683.

[52]   Zorzi, M., and R. R. Rao, "Performance of ARQ Go-Back-N Protocol in Markov Channels with Unreliable Feedback," *Mobile Network and Applications*, Vol. 2, 1997, pp. 183–193.

[53]   Nanda, S., R. Ejzac, and B. T. Doshi, "A Retransmission Scheme for Circuit—Mode Data on Wireless Links," *IEEE Journal on Selected Areas on Communication*, Vol. 12, Oct. 1994, pp. 1338–1352.

[54]   Hara, S., et al., "Throughput Performance of SAW-ARQ Protocol with Adaptive Packet Length in Mobile Data Transmission," *IEEE Trans. Veh. Tech.*, Aug. 1996, pp. 561–569.

[55]   Mandyam, G. D., "Power Control Based on Radio Link Protocol in cdma2000," *Wireless Commun. and Networking Conf.*, 1999, pp. 1368–1372.

[56]   Cianca, E., et al., "Power Management in IP-based Data Transmission over CDMA Wireless Links," in *Proc. WPMC00*, Bangkok, Nov. 2000.

[57] Kato, O., and K. Homma, "Overview on Technical Issues for Future Mobile Network and a Proposal of DLC Schemes in Wireless Link," *Special Issues on the Future Strategy for New Millennium Wireless World, Wireless Personal Communications, An International Journal,* Vol. 17, No. 2–3, June 2001, pp. 250–267.

[58] Nakamura, H., H. Matsuki, and K. Takanashi, "Efficiency of Pre-Repeat SR ARQ with Error Prediction," *IEICE Soc. Conf.,* B-307, 1996.

[59] Lambrette, U., L. Bruhl, and H. Meyr, "ARQ Protocol Performance for a Wireless High Data Rate Link," *Proc. IEEE VTC'97,* May 1997, pp. 1538–1542.

5

Adaptive Technologies

5.1 Introduction

Fast adaptation, at every level and for every possible resource, across multiple layers, is needed to meet the demand of higher data rates in the variable channel conditions typical of a wireless environment [1, 2], in order to carefully exploit the limited available resources. Adaptivity in communications networks could take different forms: *channel adaptivity* and *QoS adaptivity*.

Channel adaptivity is the ability of the network to adapt to variations in channel propagation, traffic conditions, and network topologies (for instance, in ad hoc networks) [3]. *QoS adaptivity* is the ability of the network to respond to different and varying QoS requirements of multimedia communication, such as voice, data, video, and Web.

The proper design of MAC layer protocols, and network and transport layer protocols that provide adaptive QoS has been already addressed in previous chapters of the book. This chapter will focus on channel adaptivity: the overview is necessarily nonexhaustive, due to the very wide topic that includes adaptive modulation and coding, adaptive antennas, adaptive equalization techniques. Good insight into some adaptive techniques is provided in [4, 5].

The present chapter provides basic concepts on channel adaptivity, specifically focusing on adaptive modulation and adaptive error control mechanisms, which have been popularized in the EDGE cellular system (see Chapter 1). Some results from information theory are recalled to show the limitations of these techniques and motivate further research on the practical and design issues that have to be faced to reach performance close to theoretical limits (see, in particular, Sections 5.2 and 5.3).

Furthermore, according to the objectives of the book, Sections 5.4 and 5.5 discuss trends in the implementation and design of adaptive transceivers, highlighting the need and the meaning of a cross-layered approach.

5.1.1 Diversity and Adaptation Techniques

In contrast to the additive white Gaussian noise (AWGN) channel characterized by a constant signal-to-noise ratio (SNR), in a wireless channel the SNR is time-variant due to multipath and interference from other users. There are basically two ways to utilize fully the channel capacity in presence of multipath fading and interference: adaptation and diversity techniques. In the adaptation mode, transmission parameters, such as transmission power [6], symbol transmission rate [7], constellation size [8–10], coding rate/scheme [11], or any combination of these parameters [12, 13], are changed in response to time-varying channel conditions. Diversity techniques take advantage of channel or interference level variations by resolving several fully or partially decorrelated fading channels. Performance of both techniques depends on the channel correlation. A parameter that is usually adopted to characterize the autocorrelation of a flat fading channel without line-of-sight is the Doppler frequency f_d. High values of f_d correspond to less correlated channels where significant variations of the fading coefficients over short time periods occur; low values of f_d corresponds to highly correlated channels.

In Figure 5.1 the schematic of an adaptive transmission system is shown. In order to react appropriately to channel changes, adaptation techniques require:

- A reliable prediction of the channel quality during the next active timeslots;

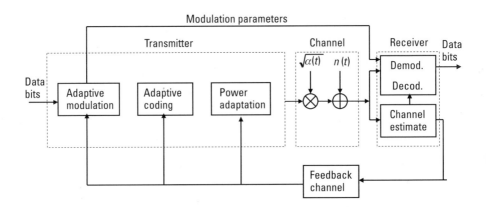

Figure 5.1 Adaptive transmission model.

- A feedback receiver-transmitter channel, in order to send back to the transmitter the channel side information (CSI);
- Information to the receiver about the demodulator parameters, which are needed to demodulate correctly the received packet;
- Flexible and fast reconfigurable transmitter and receiver terminals.

The potential of adaptive transmission was recognized 30 years ago by Cavers [7] but it did not receive much interest at that time, perhaps due to hardware constraints, lack of either good channel estimation techniques and systems with point-to-point links using transmitter feedback. During the last decade, several factors have contributed to renew the interest in adaptive techniques:

- The growing demand for spectrally efficient communication;
- The advent of feasible software radio systems and hence the availability of fast flexible and reconfigurable transceivers;
- Improvements in the prediction techniques.

Several diversity techniques have been developed, which exploit the time, frequency, or space diversity of the channel. Channel coding is a form of diversity technique. The achievable capacity gains with diversity techniques increase as the correlation among the signals exploited for the diversity reception decreases. On the other hand, successful adaptive techniques require that fast fading channel changes slowly if compared to a number of symbol periods, which corresponds to low values of the Doppler frequency f_d. If this condition is not met, the prediction information soon becomes outdated. As a consequence, the performance of these techniques degrades unless frequent transmission of quality control information is performed at the price of a significant increase of the system bandwidth requirements. Moreover, results from information theory [14–16] have shown that although the optimal adaptive technique that exploits the CSI at both transmitter and receiver always has the highest capacity with respect to scheme that exploits the CSI only at the receiver, this capacity gain becomes small when the fading is approximately i.i.d. This result is independent of the effects of estimation errors and delays.

In Figure 5.2, the performance trend of diversity techniques and adaptation techniques in terms of signal quality are shown as a function of the channel Doppler frequency.

As a matter of fact, a high diversity order converts a fading channel into an AWGN-like channel, and adaptation to the channel variations (through adaptive modulation or coding) is unnecessary. An increase of the diversity order requires an increase of the receiver and transmitter hardware and involves a

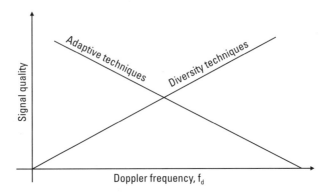

Figure 5.2 Behavior of adaptive techniques and diversity techniques as a function of the channel correlation.

higher computational complexity (e.g., a high number of antennas for space-time diversity techniques). Adaptive modulation can be seen as a lower-complexity alternative to diversity techniques since a single transmitter and receiver are required [4]. On the other hand, adaptive techniques that utilize CSI at the transmitter increase encoding and decoding complexity. Therefore, the trade-off between adaptive and nonadaptive techniques is both on the capacity and the complexity.

5.1.2   Modulation and Channel Coding

This section introduces the basics concepts and definitions on modulation and coding in order to facilitate the reading of the rest of the chapter.

The modulation process maps the digital information into analog waveforms that match the characteristics of the channel. Blocks of $k = \log_2 M$ binary symbols of the information sequence $\{a_n\}$ are generally associated with one of $M = 2^k$ deterministic, finite energy waveforms $\{s_m(t), m = 1, 2..., M\}$ for transmission over the channel. These waveforms may differ in either amplitude or phase or frequency, as well as in some combination of two or more signal parameters. Signals waveforms corresponding to carrier amplitude modulation (AM) and phase modulation (PM) or a combination of them may be represented as:

$$s_m(t) = \mathrm{Re}\left[e^{j(2\pi f_0 t + \varphi_0)} \sum_{n=-\infty}^{+\infty} c_n g_T(t - nT) \right] \qquad (5.1)$$

where $\mathrm{Re}[\cdot]$ takes the real-part of the complex signal, f_0 and φ_0 denote the carrier frequency and phase, $T = kT_b$ is the symbol interval (T_b is the bit interval), $g_T(\cdot)$ is a real-valued signal pulse that represents the transmission filter and c_n are generally complex-valued symbols belonging to an alphabet of size M. This alphabet is also called signal *constellation*. In Figure 5.3(a), the constellation symbols of two pulse amplitude modulation (PAM) signals are shown.

The signal space diagrams for phase shift keying (PSK) signals are shown in Figure 5.3(b). In this case, signal waveforms differ only for carrier phase. Amplitude modulation in two orthogonal phase directions are used in the quadrature amplitude modulation (QAM) schemes as shown in Figure 5.3(c). In all these schemes the assignment of k information bits to the M possible symbol c_n is usually done in such a way that adjacent symbols differ by one binary digit. This mapping is called *Grey encoding* and ensures that most likely errors caused by noise will result in a single bit error in the k-bit symbol. With this assignment, the minimum Euclidean distance (i.e., the distance between a pair of adjacent signal points) is

$$d^e_{min} = d\sqrt{2E_g} \qquad \text{for a } M \text{ PAM and rectangular } M \text{ QAM}$$

$$d^e_{min} = d\sqrt{E_g\left(1 - \cos\frac{2\pi}{M}\right)} \quad \text{for a } M \text{ PSK}$$

$$(5.2)$$

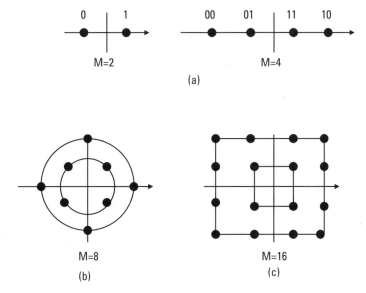

Figure 5.3 Signal space diagrams for digital modulated signals: (a) PAM constellation; (b) 8-PSK constellation; and (c) rectangular 16-QAM constellation.

where $2d$ is the distance between adjacent signal amplitudes and E_g is the energy in the pulse $g_T(\cdot)$. A higher-order constellation is characterized by lower $d^e{}_{min}$, if the average transmitted power per symbol is kept constant. Therefore, let us denote with SIR_{thr} the minimum signal-to-interference ratio that allows one to achieve a BER lower than a target value. Binary and quaternary modulation schemes, which only encode 1 or 2 bits per symbol, are characterized by lower SIR_{thr} with respect to higher-level modulation schemes (M-QAM), which encode more bits per symbol.

The modulation scheme impacts spectrum efficiency in two ways, as it is clarified by the following example. Let us consider a cellular system with a hexagonal cell structure that has six significant interferers, regardless the cluster size [17].

The SIR is the received signal power from the desired BS divided by the summation of the interference powers from cochannel BSs. Based on the logarithmic path loss model, neglecting noise and assuming that the signal level from all the interferers is approximately equal, the SIR becomes

$$SIR = \frac{R^{-\gamma}}{\left(6D^{-\gamma}\right)} \tag{5.3}$$

where R is the radius of a cell, D the distance between interfering cells, and the path loss exponent. As

$$D/R = \sqrt{3K} \tag{5.4}$$

where K is the number of cells in a frequency reuse pattern (depending on the SIR_{thr}), the number, m, of radio channels available in each cell is

$$m\left(M_i\right) = \frac{B_t}{B_c\left(M_i\right)K} \tag{5.5}$$

where B_t is the total bandwidth available and $B_c(M_i)$ is the bandwidth required per channel and for the modulation scheme M_i. Assuming $\gamma = 4$, the number of available channels in each cell is

$$m\left(M_i\right) = \frac{B_t}{B_c\left(M_i\right)\sqrt{\dfrac{2}{3}SIR_{thr}\left(M_i\right)}} \tag{5.6}$$

By choosing a higher-level modulation scheme, B_c turns out to be reduced but SIR_{thr} is also higher. The latter trade-off must be considered in order to select the modulation scheme that provides a higher spectrum efficiency for a given propagation scenario.

Channel coding consists of adding redundancy bits to the transmitted bit stream to allow the receiver to detect, and in some cases to correct, transmission errors. The encoder associates to an input word of k symbols a corresponding output word composed of a larger number of symbols n. The ratio k/n is known as code rate R_c. The lower R_c, the more robust the coding. The increased robustness to the transmission errors is paid for by a reduction in the effective bit rate, given the same average transmission power.

Many algorithms can be used to obtain the output words from the input ones of the encoder. Two main classes of algorithms can be defined: block codes and tree codes. Block codes need no memory of previous input data words in the encoder and, hence, the output word depends only on the current input data word. Examples of block codes are the Bose-Chaudhuri-Hocquenghem (BCH) codes, Golay codes, and Reed Solomon codes. In particular, Reed Solomon codes group together several bits into multilevel symbols before encoding. In this way they are able to correct error bursts. Tree codes use previous input words to encode the current input word. In Figure 5.4, a tree code encoder with shift register is shown.

The incoming information sequence is broken into segments of k_0 symbols, called information frames. The encoder can store m frames. From the incoming information frame and the m stored frames, the encoder computes a single codeword frame of length n_0 symbols. This codeword is shifted out of the encoder as the next information frame is shifted in. During each frame time, a new information frame is shifted into the shift register, and the oldest

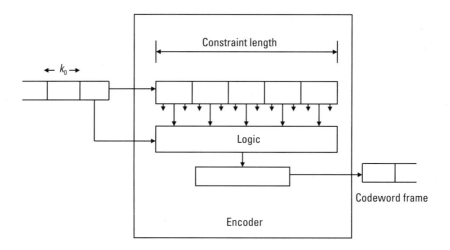

Figure 5.4 Tree code shift register encoder.

information frame is shifted out and discarded. The parameter $K = mk_0$ is called constraints length. The constraint length can be finite or infinite. Practical tree codes have finite constraints lengths and are called *trellis codes*. If the tree has finite memory and is linear, the tree code is a *convolutional code*.

5.2   Adaptive Modulation

Variable-rate modulation systems vary the number of modulation levels (i.e., the size of the constellation alphabet) according to a quality criterion. The development of adaptive modulation systems requires the specification of:

1. The selection rule of the modulation parameters;
2. Channel estimation techniques;
3. Modulation parameter estimation at the receiver (to enable a correct demodulation).

5.2.1   Modulation Parameter Selection

Several modulation level–controlled adaptive modulation systems have been proposed where modulation level is selected according to the traffic or received signal level [8, 18].

Figure 5.5 shows an illustrative example where the modulation level is chosen from the estimation of the instantaneous SNR and an average symbol error probability is aimed.

The modulation mode is selected according the following switching algorithm:

$$\text{Modutation mode} = \begin{cases} NOTX & \text{if} & SNR \leq thr1 \\ BPSK & \text{if} & thr1 < SNR \leq thr2 \\ 4QAM & \text{if} & thr2 < SNR \leq thr3 \\ 16QAM & \text{if} & thr3 < SNR \leq thr4 \\ 64QAM & \text{if} & SNR > thr4 \end{cases}$$

According to this modulation switching algorithm, 64-QAM is used as the maximum modulation level, thus transmitting 6 bits per symbol when the channel is at its best (indoor scenarios, center of a cell), while the robust but less spectrally efficient BPSK is adopted when both channels degrade. The numerical upper bound performance of narrowband adaptive QAM was evaluated over slow Rayleigh flat-fading channels in [19], and over wideband channels in [20, 21]. Optimized switching thresholds for adaptive QAM have been carried out

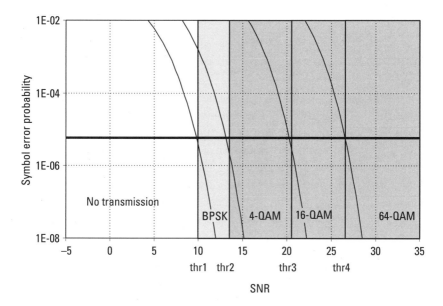

Figure 5.5 Modulation level selection based on the symbol error rate.

employing Powell-optimization using a cost function that was based on a combination of target BER and target bit per symbol (BPS) performance [22].

Modulation level–controlled adaptive modulation systems are characterized by the following specific drawbacks [9]:

1. The modulation parameter control range is not sufficient to cover signal level variations due to fading. For example, the controlled range of the adaptive modulation using QPSK to 256-QAM has only about 15 dB.

2. In frequency-selective fading environments, modulation level control cannot adequately improve delay-spread immunity because delay-spread immunity is closely related to the symbol rate.

3. Variable QAM levels in response to fading conditions result in variable bit rate, which although nearly constant over long periods, could instantaneously vary by several times the average rate.

To achieve a wider modulation parameter control range, other signal parameters, together with the modulation level, can be varied according to the channel conditions. In [9], this wider control range together with an improvement in the delay-spread immunity is achieved by a scheme that controls the symbol rate together with the modulation level. The predicted channel parameters used to drive the control are the carrier-to-noise power ratio C/N_0 and the delay spread: both parameters are calculated by the delay profile that is predicted by the previous timeslot at the receiver. In this particular channel monitoring technique, the channel reciprocity between the uplink and the dowlink is exploited since a TDD mode is assumed and the time interval between the transmission and reception timeslot is sufficiently short to assume high correlation between the two delay profiles. The considered system selects higher symbol rate and higher modulation level when the C/N_0 is high and the delay spread is small. On the other hand, when C/N_0 is low or delay spread is large, lower symbol rate and/or lower modulation level are selected. In particular, when the delay spread is large, performance could be improved by reducing the symbol rate in such a way that no other techniques are needed to cope with frequency selective fading.

The optimal adaptive transmission scheme that achieves the Shannon capacity of a fading channel was derived in [14]. In the treatise, a slowly varying channel was assumed and the instantaneous received power required to achieve a certain upper bound performance was assumed to be known prior to transmission.

Moving from the above result, a practical adaptive modulation technique for fading channels that controls both the modulation level and the transmission power is presented in [23]. Let $S(\gamma)$ denote the transmit power adaptation policy relative to an instantaneous value γ for the SNR. It is subjected to the average power constraint:

$$\int_0^\infty S(\gamma)p(\gamma)d\gamma \leq \overline{S} \tag{5.7}$$

The power adaptation policy, which maximizes the spectral efficiency, is such that

$$\frac{S(\gamma)}{\overline{S}} = \begin{cases} \dfrac{1}{\gamma_0} - \dfrac{1}{\gamma\Gamma}, & \gamma \geq \gamma_0/\Gamma \\ 0 & \gamma < \gamma_0/\Gamma \end{cases} \tag{5.8}$$

where γ_0 is the cutoff fade depth of the optimal transmission scheme on fading channel, and γ_0/Γ is the optimized cutoff fade depth of the practical adaptive modulation technique, where

$$\Gamma = \frac{-1.5}{\ln(5BER)} \tag{5.9}$$

The achieved maximum spectral efficiency is

$$\frac{R}{B} = \int_{\gamma_{0/K}}^{\infty} \log_2\left(\frac{\gamma\Gamma}{\gamma_0}\right) p(\gamma)d\gamma \tag{5.10}$$

The maximum spectral efficiency of an optimal transmission scheme on fading channel is given by [14]

$$\frac{R}{B} = \int_{\gamma_0}^{\infty} \log_2\left(\frac{\gamma}{\gamma_0}\right) p(\gamma)d\gamma \tag{5.11}$$

This adaptive technique has shown [23] a 5- to 10-dB power gain over a variable-power fixed rate modulation, and up to 20 dB of power gain over nonadaptive modulation.

The technique, however, is sensitive to channel estimation errors and to estimation and feedback path delay, and this must be taken into account in any practical implementation. The comparison between (5.10) and (5.11) shows that regardless the fading distribution, the maximum possible coding gain for the identified adaptive uncoded MQAM scheme is K. The exploitation of coding with adaptive modulation is discussed in Section 5.2.4.

The fact that variable QAM levels in response to fading conditions result in variable bit rate is not a limitation in the context of data transmission, but it does lead to the need of appropriate source codecs that are capable of promptly reconfiguring themselves and suitable buffering arrangements in interactive speech or video communications over a modem that exploit adaptive modulation, thus increasing the delay experienced by the user [10]. In [24, 25], the latency associated with storing the information to be transmitted during severally degraded channel conditions was mitigated by frequency hopping or statistical multiplexing, thus achieving more than 4-dB SNR reduction with respect to conventional nonadaptive modems. The achievable gains, however, were strongly affected by the cochannel interference level. Interference cancellation may be adopted [25], thus adapting the demodulation decision boundaries after estimating the interfering channel magnitude and phase. Alternatively, power adaptation combined with fixed-rate transmission might be a suitable solution for voice transmission, which has low data rate requirements with real-time delay constraints. In [26], a new adaptive modulation scheme is proposed for simultaneous voice and data transmission over fading channels. A fixed-rate BPSK modulation on the quadrature (Q) channel for voice communication is adopted, and variable-rate M-ary amplitude (M-AM) modulation is adopted on the in-phase (I) channel for data. In case of bad channels, priority is given to voice communication by allocating most of the transmitted power to ensure continuous and satisfactory transmission of speech communications. If the power required to meet this target exceeds a peak power constraint, a voice outage is declared. The remaining power is used for data communications on the I channel, where adaptive modulation is used to meet the target BER. As channel conditions improve, most of the transmitted power is reallocated to high data rate transmission on the I channel. This technique has shown much lower spectral efficiency with respect to other adaptive modulation schemes. Other solutions have been investigated and are still under investigation for systems that simultaneously meet the BER and delay constraints of voice and data. One candidate for this adaptive modulation uses unequal error protection (UEP) signal constellation [27].

5.2.2   Channel Prediction

Adaptive modulation requires accurate channel prediction at the receiver, which is fed back to the transmitter with minimal latency. Feedback delay and overhead, processing delay, and practical constraints on modulation, coding, and/or antenna switching rates have to be taken into account in the performance analysis of adaptive transmission methods. The high carrier frequency of third-generation mobile systems (2 GHz) will result in a very large Doppler shifts at moderate vehicular speeds (e.g., 65 mi/h, $f_d = 200$ Hz) that cause significant variations of the fading channel coefficients over short time periods. Therefore, even a small delay will cause significant degradation of performance since channel variation, due to large Doppler shifts, usually results in a different channel at the time of transmission than at the time of channel estimation. The effect of estimation error and feedback delay on adaptive modulation were analyzed in [23], where it was found that, for a target BER of 10^{-6}, the BER remains at its target level as long as the total delay of the channel estimator and feedback path is less than $0.001 \lambda/v$ where v is the vehicle speed and λ is the signal wavelength. Moreover, assuming perfect automatic gain control (AGC), the estimation error must be less than 1 dB to maintain the target BER. This bound on the estimation error variance can be achieved by a pilot symbol assisted technique [28] where the transmitter periodically insert known symbols that are exploited at the receiver for channel estimation. The pilot symbols lower the effective bit rate depending on the Doppler frequency of the channel.

Many research works have addressed the problem of the estimation of current fading conditions [29–31]. In [9] and [32], the problem of the prediction is addressed but either short-range prediction or very slow fading is considered in these investigations, while the accurate several tens-to-hundreds symbols ahead prediction of channel coefficients is essential for implementing the potential of adaptive transmission [31]. Several proposed long-range prediction channel methods require a long observation interval and heavy computational load to compute current model parameters and they are not suitable in vary fast varying channels. Therefore, the effort is in providing low-complexity long-range prediction techniques [31].

5.2.3 Modulation Parameter Estimation at the Receiver

As for the modulation parameter estimation, in [9] an estimation word is embedded in the midamble of each timeslot. The word consists of an eight-symbol Walsh function. Seven options of modulation parameters are considered and hence, seven out of eight different Walsh codes are employed, each corresponding to a set of modulation parameters. At the receiver, correlation between the received word and all the codeword candidates is taken, and the codeword having the maximum correlation value is assumed to be the transmitted codeword. Another technique of estimating the required modulation mode was proposed in [33], where the modulation control symbols were represented by unequal error protection 5-PSK symbols. When the information is conveyed to the receiver, there is a loss of effective data throughput. Alternatively, the receiver can attempt to estimate the parameters employed by the remote transmitter by means of blind detection mechanisms.

5.2.4 Coding in Adaptive Modulation Schemes

Codes designed for AWGN channel can be superimposed on a very general class of adaptive modulation schemes, with the same approximate coding gain [34]. However, block and convolutional codes are not spectrally efficient, and would thus reduce some of the efficiency gains of the variable-rate scheme. More effective coding schemes are trellis and lattice codes, which are special cases of coset codes. For these codes, the code design and the modulation design are separable [35, 36], and hence, size, power, and symbol rate can be varied without affecting BER or coding gain.

A simple four-trellis code yields an asymptotic coding gain of 3 dB and an eight-state code yields an asymptotic coding gain of 4 dB [34]. Furthermore, it has been found that is difficult to obtain more than 4 dB of coding gain using a trellis code of reasonable complexity. Turbo codes [37–42] do yield higher gains, but the design and analysis is more complicated. Coding and modulation design cannot be separated in case of Turbo codes; hence, it is not clear if they achieve the same coding gain as in AWGN channels when superimposed onto adaptive modulation. Therefore, the gap between the spectral efficiency of adaptive modulation and Shannon capacity, already highlighted, cannot be fully closed. This is a consequence of the lack of complexity and implementation constraints inherent to Shannon theory.

Moreover, adaptive coded modulation does not require interleaving. In fact, if the adaptive modulation keeps the BER constant under all fading conditions—by adjusting the transmit power and rate—the probability of errors in a deep fade is the same as with little or no fading, thereby eliminating error bursts.

5.3   Adaptive Error Control

There are two main error control procedures: FEC and ARQ [43]. The reader may refer to Chapter 4 for a description of basic ARQ schemes and Section 5.1.2 for the basics on channel coding. Instead of fixing a level of overhead that can cope with worst-case conditions, adaptive error control mechanisms let the error protection vary as the conditions vary. That is, the overhead is always adapted to the current conditions, avoiding both over-pessimistic channel coding when conditions are good, and resource wasting retransmissions due to insufficient error protection.

In the following sections, the adaptive implementation of the error control mechanisms shown in Table 5.1 is presented.

5.3.1   Adaptive FEC

Adaptive FEC is performed by changing code rates [44]. For practical purposes, it is desirable to modify code rates without changing the basic encoder and decoder structure. Punctured convolutional codes are well suited for this application since they provide a wide range of code rates by using the same encoder and the same maximum likelihood decoder with Viterbi algorithm. A punctured convolutional code is a high-rate code obtained by periodic elimination (i.e., puncturing) of specific code symbols from a low-rate code [45]. The pattern of punctured symbols is called the perforation partner of the punctured code and it is conventionally described in a matrix called *perforation matrix*. Let us consider a low-rate code of R_{low}. By deleting $S = (b/R_{low}-v)$ symbols from every b/R_{low} code symbols, the desired code rate $R_c = b/v$ is achieved. Variable-rate coding can be readily obtained if all punctured rates of interest are obtained from the same low-rate encoder. Only the *perforation matrix* has to be modified accordingly. At the receiver end, the Viterbi decoder operates on the trellis of the root low-rate code and uses the same deleting map as in the encoder in computing path metrics [45]. Therefore, Viterbi codecs for high-rate punctured convolutional codes have not the same complexity of a straightforward decoding of a

Table 5.1
Adaptive Error Control Mechanisms

Adaptive Error Control Schemes		
Adaptive FEC	Hybrid ARQ type I, type II, type III	Adaptive ARQ

high-rate *b/v* code and they can be implemented by adding relatively simple hardware to the codes of the original low-rate codec.

Therefore, implementation of adaptive coding can be achieved by a modest increase in hardware. It is worth noting that there is no difference with fixed-rate and variable-rate coding schemes in terms of channel capacity [46]. On the other hand, the error exponent, which describes how fast error probability drops with respect to block length, is significantly increased using adaptive coding rate control. Recent work [23, 47] has shown that significant gains in bandwidth efficiency are exhibited by adaptive trellis-coded modulation (TCM) [48] schemes over their nonadaptive counterparts for delay-constrained, low-complexity decoders. Hence, while variable-rate coding may not significantly increase capacity, they may provide better performance for low-complexity or delay-constrained systems.

5.3.2 Hybrid ARQ

Hybrid ARQ (HARQ) schemes combine FEC and ARQ in different ways. A pure ARQ scheme may provide a higher throughput (received correct bits/second) in good channel conditions since the redundancy bits added in each transmission consist only of CRC bits. An FEC code has better performance when the channel conditions are bad since the higher redundancy added by the FEC code is needed most of the time to correctly recover the transmitted data. A combination of FEC and ARQ can perform better than pure FEC or pure ARQ in time-varying channels. When the channel conditions are bad, the FEC code helps in reducing the probability of retransmitted packet loss, while when the channel is good, the retransmission of erroneously received packets may provide a good reliability with a lower level of redundancy than a pure FEC error control.

Three types of HARQ schemes may be identified.

Type I HARQ

The same frame is transmitted in all retransmissions. At the receiver, diversity combining or code combining may be used to improve reliability [49]. Therefore, packets that are not correctly received are not immediately discarded but may be used to decode the packet. Two configurations are possible:

1. Data + FEC overhead;
2. (Data + CRC) + FEC.

In the first configuration, when the number of errors in the received block exceeds the code correcting capability, the FEC is used as an error detection code. In the second and more reliable configuration, the decoded sequence consists of information data and error detection code.

Type II HARQ

In this configuration, only parity bits are sent in each retransmission. Most type II algorithms adopt an incremental parity retransmission scheme [50–52], which uses an invertible code. A code is said to be invertible if, knowing only the parity check bits of a code word, the corresponding information can be uniquely determined by an inversion process. Efficient type II HARQ schemes can be achieved by adopting rate-compatible convolutional codes (RCPC) [53]. The rate compatibility condition insures that all coded bits of any code of the family are used by all lower-rate codes. Therefore, if an invertible rate $1/m$ code is assumed, at the first retransmission only the bits corresponding to the higher-rate code of the family are sent. Next, blocks of parity bits are sent in subsequent retransmissions and are combined at the receiver with the previous ones, to recover the data. After m retransmissions, if the received word is still assumed erroneous, the original word is sent again in the subsequent retransmission.

The block diagram of a type II HARQ algorithm that uses an invertible half-rate code for the error correction is shown in Figure 5.6.

In the protocol described in Figure 5.6, the process continues until the data is correctly retrieved. For an ARQ scheme with a maximum number of

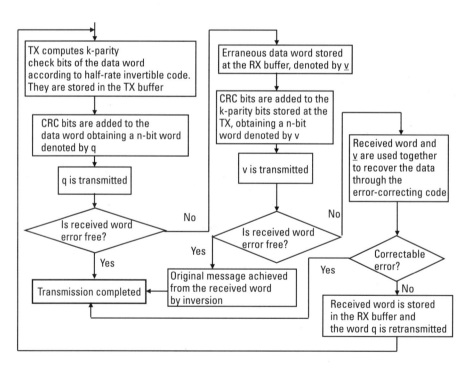

Figure 5.6 Flow diagram of an HARQ type II with infinite persistency and an invertible half-rate code (TX: transmitter, RX: receiver).

retransmissions n_{max}, failure is declared when an erroneous word is received after n_{max} transmission attempts.

Turbo codes have been also proposed as candidates for packet combining since they are systematic and produce incremental redundancy by puncturing parity bits [54, 55].

Type III HARQ

Both user data and parity bits are included in every retransmission. They are based on complementary punctured convolutional (CPC) codes [56]. A set of punctured convolutional codes derived from the same original low-rate code are said to be complementary if they are equivalent (in terms of their distance properties) and their combination yields at least the original low-rate code. The block diagram of a type III HARQ scheme is shown in Figure 5.7, where $n_{max} = 1$ is assumed.

In general, at the ith transmission attempt, Viterbi is first applied on the received word, using the perforation pattern P_i. If the decoded sequence is assumed error free, transmission is completed. Otherwise, Viterbi decoding is applied once again but using the combined code that has perforation pattern $P^{(i)} = P_1 + P_2 + \dots P_i$.

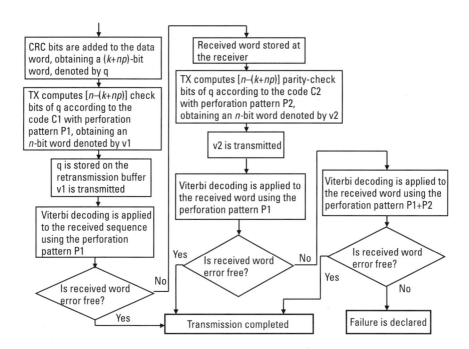

Figure 5.7 Flow diagram of an HARQ type III with finite persistency. Maximum number of retransmissions is 1.

A summary of the main features of the presented HARQ schemes is shown in Table 5.2.

The amount of FEC overhead is fixed in the case of type I HARQ, and it is higher than in a pure ARQ scheme. As a result, when the BER is low, the corresponding ARQ scheme has a higher throughput than the hybrid one. In both type II and type III HARQ, the overhead due to the FEC and retransmissions depends on the channel conditions. Therefore, type I HARQ schemes are not well suited for time-varying channels, while both type II and type III HARQ schemes provide adaptation to channel conditions. In good channel conditions, a minimum level of redundancy is added. Specifically, in type II schemes the redundancy is only due to the error detection coding (like in a pure ARQ), while in type III schemes it is due to the error detection plus an error correcting coding that matches the channel noise requirements. In case of bad channel conditions, more powerful codes are used in the decoding process by combining previously received erroneous sequences. In fact, in both type III and type II HARQ schemes, erroneously received words are not discarded but stored at the receiver and combined for decoding. However, note that in type II HARQ the previously received sequence is essential for the decoding process, while in type

Table 5.2
Comparison of HARQ Schemes

HARQ Scheme	Type I	Type II	Type III
Description	Same frame sent in all the retransmissions	Only parity bits sent in some retransmissions	User data and parity bits are included in each retransmission
	Packets detected in error are discarded	Packets detected in error are not discarded	Packets detected in error are not discarded
	User data may be recovered from each single transmission	Decoder has to rely on previously received word for the same data packet	User data may be recovered from each single transmission
Coding overhead at the transmission	High	Low	Medium
Suitability	No time-varying channels	Time-varying channels	Time-varying channels. Better than type II in bursty channels
	Better than pure ARQ in good channel conditions		

RTX: retransmission.

III schemes user data may be recovered from each single transmission. This property of type III HARQ scheme is called *self-decodability* and it is very useful in bursty channels, where several consecutive packets may be damaged.

Information theory results [57] have shown that the maximum throughput of an incremental redundancy scheme based on progressively punctured codes (INR) is achieved for infinite delay. The same maximum throughput (with zero packet loss probability) can be achieved by a system without feedback (just forward error correction) with infinite delay [58]. The need of implementing error control with feedback is due to the fact that these schemes can achieve a zero transmission failure probability with finite average delay for all the rates strictly less than the capacity of the channel, while systems without feedback need a very large (infinite) delay for all values of throughput.

5.3.3 Adaptive ARQ

Let us define the sequence of transmissions or attempts that are used for delivering a block of user bits. For pure FEC schemes a cycle consists of a single transmission, while for ARQ schemes, a cycle may consist of more than one transmission. Hybrid ARQ are always able to adapt the user information rate within the cycle to the channel state (or to the average channel state) [59]. However, an ARQ scheme is *adaptive* in the case that it is able to adapt the user information rate of the first transmission within a cycle [60–62]. Some adaptive ARQ schemes are now considered.

The efficiency of an ARQ protocol for delivering useful data can be measured by the throughput, defined as the ratio between the average number of bits accepted by the receiver per unit time and the total number of bits that could be sent on the channel per unit time. The throughput of a selective repeat (SR) scheme (see Chapter 4 for an introduction on ARQ schemes) is given by the well-known formula [43]:

$$\eta = \frac{n-h}{n}\left(1 - p_e\right)^n \qquad (5.12)$$

where n is the packet size in bits, h is the number of overhead bits (including CRC); and p_e is the channel BER.

In Figure 5.8, the throughput curves for the SR ARQ protocol are plotted as a function of the block size and different values of the BER.

Figure 5.8 shows that for each target BER there is an optimum value of the block size that maximizes the throughput. As expected, as the BER increases, the throughput can be improved by choosing a lower block size. The adaptive ARQ scheme proposed in [62] uses a look-up table with a number of rows equal to the number of switching points. To each switching point corresponds a BER and a

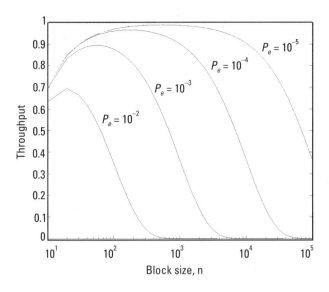

Figure 5.8 Throughput of an SR-ARQ protocol versus block size for different values of the error probability.

block size. When the BER is between the BER value associated to the switching point i and $i + 1$, the block size corresponding to the switching point i is chosen. This procedure requires an estimate of the exact BER, which is performed as

$$BER = \frac{TNBE}{TNB \cdot n}$$ (5.13)

where $TNBE$ is the number of n-bit packets in error, and TNB is the total number of n-bit packets since the time that the packet size was changed last. The main disadvantage of this scheme is that the time interval observation interval (OBI), over which the estimate is computed, becomes arbitrarily long.

Another approach consists of varying the number of contiguously retransmitted copies of the erroneous block. By sending contiguously more than one copy of the erroneous block, the efficiency of the algorithm can be improved for packet error rate above a certain threshold, which in turn depends on the round trip [63]. In [64], the number of contiguously transmitted copies in a GBN algorithm is determined by using the knowledge of the block error rate. In a slowly varying channel, the optimal choice for a multicopy GBN performs as an $m1$-copy GBN scheme when the packet error rate (PER) is low ($m1$ is the number of contiguously retransmitted copies of the erroneous block) and as an $m2$-copy GBN when the PER is high, with $m_1 < m_2$. In [65], a GBN adaptive scheme is proposed where $m1 = 1$ and $m2 = 2$ and the channel state is estimated by counting the number of contiguous ACKs/NACKs. Let us denote with L and H the two transmission modes for good and bad channel conditions respectively; then,

- If in the "good" state, the transmitter receives α contiguous NACKs, it considers the channel is in "bad" state, and changes to a H transmission mode;
- In the "bad" state, if β contiguous ACKs are received, the transmitter reverts to L mode.

The system can be modeled by a Markov chain with $2(\alpha + \beta)$ states. The state transition diagram of this multicopy GBN scheme, referred as the Yao's scheme, is shown in Figure 5.9, with $\alpha = 2$ and $\beta = 3$ [60].

In Figure 5.9, in the GL_i states ($i = 0, ..., \alpha - 1$), the channel is good (G), the transmission mode is L, and the transmitter has received i contiguous NACKs; in the GH_i states ($i = 0, ..., \beta - 1$), the channel is good, the transmission mode is H and the transmitter has received i contiguous ACKs; in the BL_i states ($i = 0, ..., \alpha - 1$), the channel is bad, the transmission mode is L and the transmitter has received i contiguous NACKs; in the BH_i states ($i = 0, ..., \beta - 1$), the channel is bad, the transmission mode is H and the transmitter has received i contiguous ACKs. The channel sensing algorithm turns out to be a

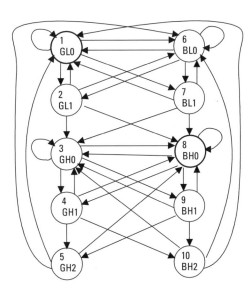

Figure 5.9 State transition diagram of the Markov model in Yao's multicopy GBN scheme with $\alpha = 2$ and $\beta = 3$.

sliding OBI algorithm, with sliding intervals of lengths α and β, which has a shorter response time to the channel changes with respect to not sliding OBI algorithms. Furthermore, note that α and β were heuristically chosen in [65]. An optimization of these parameters has been tried in [66], where it has been concluded that only suboptimal values of them can be found in a Gilbert-Elliot channel. However, the optimal choice of parameters (α, β) depends heavily on the channel model, and this is often not adequately considered in most of the throughput analyses of ARQ schemes in literature [67].

Another approach for adaptive ARQ error control is proposed in [68], where Rice and Wicker suggested a code rate adaptive error control over a Gilbert-Elliot channel with three binary symmetric channel (BSC) error states (low, medium, high), as follows:

1. In ARQ scheme with (127,106) BCH code, for low state;
2. Hybrid-ARQ scheme with (127,99) BCH code, for medium state;
3. FEC scheme with (127,78) BCH code, for high state.

With a guarantee of undetected errors lower than 10^{-6}, up to a BER of 10^{-1}.

In Table 5.3, the three introduced approaches of adaptive ARQ are compared in terms of reaction capability to the channel state variations. This capability is mainly due to the channel state monitor strategy since an adaptive ARQ scheme can react only when the channel state has been estimated. A type II hybrid ARQ scheme can react to the channel changes more quickly as it sends parity bits in the first retransmission. However, it is still an open issue if an optimized adaptive SR-ARQ can be more efficient than a type II HARQ. Moreover, hybrid ARQ schemes typically assume fixed block sizes. It is expected that type II hybrid ARQ could benefit from adaptive strategies such as variable packet size variation.

5.4 Multilayer Adaptivity

Previous sections of the chapter focus on adaptation techniques at the physical and data link layer, which basically provide adaptation to the available communications resources, including radio channel. To take full advantage of the adaptivity, a pure-layered protocol and architectures design approach is insufficient; rather, a cross-layered design approach is needed, where information among layers is exchanged and parameters are jointly optimized. In Chapter 4, an example of a cross-layered design approach has been presented when design issues of some physical and data link layer protocols have been addressed in order to optimize end-to-end performance at transport layer. Figure 5.10 highlights the interaction among layers and the multilayer functionality of the adaptive resource management in a fully adaptive architecture [2].

Table 5.3
Reaction Capability of Some Adaptive ARQ Schemes

Adaptive ARQ Approach	Channel State Monitor Strategy	Reaction Capability to Channel State Changes
1. Varying block size	Estimate of the BER from (5.13)	Slow, because the OBI algorithm is not slidiing window
2. Varying the number of contiguously retransmitted copies	Number of contiguous ACKs/NACKs	Higher than in approach 1, since the OBI algorithm is sliding window
3. Switching between different ARQ/FEC schemes		Slower than hybrid ARQ

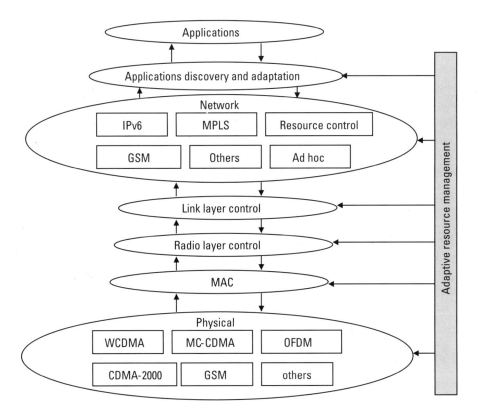

Figure 5.10 Adaptive multilayer architecture.

The adaptive resource management includes activities decision and control, adjusting parameters and functions in order to optimize desired features such as QoS, throughput, power utilization, or possibly an overall cost function.

In both 2G and 3G standards, the RRM function has already a multilayer control, from the network layer down to the data link and physical layers. However, the adaptive radio resource management in future systems will be a much more complex function. Similarly to the RRM in current 2G and 3G systems, a function of the resource management unit is to control modulation and coding functionality as well as to provide parametric control over constellation sizes, power levels, code sizes, channel measurements, and other related functions. New features of future RRM units is that they will have the capability not just to select parameters, but also to select the adaptation mechanisms, as well as the capability of varying on a faster timeframe. Chapter 6 is entirely dedicated to the resource management issue in future communication systems.

In the multilayer adaptive architecture shown in Figure 5.10, an adaptation layer is inserted between the network and application layers. One of its functions is the application discovery, which is the ability to identify useful applications that are valuable to a specific user (e.g., business, nonbusiness) or system. The discovery mechanisms allow the user to search selectively, according to its needs, and to ignore the rest. It includes negotiation of communications services, such as resources (e.g., data rates, QoS). The adaptation layer allows nonnative payloads to use transport and lower layer functionality. It provides adaptation to the available communication resources in response to channel capacity and delay conditions and yields to a negotiation degree between user need and available resources, including also pricing options.

Many challenging research issues are created by the need to build richer interfaces among the layers of the protocol stack in the future wireless systems, such as:

- The need for a proper modeling of the lower layers in order to evaluate and optimize the performance of higher layers protocols (see also Chapter 4);
- Proper definition of the set of performance metrics and parameters that serve as agents to carry the information between layers; they should be simple and robust to modeling errors.

5.5   Hardware and Software Implementations

The achievable gains of adaptive transceivers are strongly related to the flexibility offered by feasible software radio implementations of hardware devices [69–72]. The software radio concept is based on the inherent programmability of digital signal processing hardware (DSPH) such as DSP and/or FPGA and/or ASIC.

Typical software languages that are used to configure DSPH are VHDL and Verilog-HDL. Software radio systems can perform multimode modulation and demodulation on a per-packet basis [73], offering greater control over spectrum usage than dedicated hardware, and they are amenable to future software upgrades to support future signal processing algorithms or yet unknown coding. On the other hand, so far, most of the software implementations have been restricted to non-real-time operation, since finite processor power limits the complexity of a software radio if real-time constraints have to be met. The need to process signals continuously at relatively high speed and at the lowest possible cost has typically led to a selection of the lowest-cost implementation technology, as well as specific modulation and coding schemes with only a certain amount of parametric options. Moreover, hardware radios and hardware implementations are typically less expensive than software ones for large-scale production. Recently, however, a greater degree of programmability is being made available on devices that are nearly as inexpensive as fixed function devices, and software infrastructures to control these devices are emerging. A number of companies are making baseband processors available that include greater flexibility than ever before. Baseband chipsets able to carry out WCDMA and GSM, or cdma1x and AMPS, or 802.11b, WCDMA, and GSM are emerging [2]. It seems there is no near-term limit to the evolution of programmable signal and baseband processors and to the software infrastructures that will provide a platform for adaptability.

5.6 Conclusions

Adaptivity is one of the main technologies for future wireless communication multimedia systems. Information theoretical results have shown that when CSI is made available to the transmitter, adaptive techniques can greatly improve performance in systems with stringent delay constraints, without the need of interleaving or the exploitation of diversity techniques. Adaptive modulation can provide a five-fold increase in the spectral efficiency, and 3 to 6 dB of coding gain are provided by adaptive coded modulation with Trellis codes with respect to uncoded transmission. Exploiting error control mechanisms with feedback, such as ARQ, can reduce the delay needed to achieve a given performance in terms of BER. Turbo codes (or other forms of concatenated coding) with iterative decoding appear to be a promising solution for hybrid ARQ schemes. However, the behavior of iterative decoders in the presence of decoding errors should be better characterized in order to exploit it for error detection.

The gain that can be achieved by channel adaptive techniques strictly depends on the knowledge of the current channel fading value. Therefore, new algorithms for fast and reliable long-range channel prediction are under investigation. At the same time, solutions for extending the applicability of adaptive techniques to systems where prediction techniques are not effective (i.e., systems with users characterized by high mobility) are being proposed [74].

The knowledge of CSI at the transmitter requires signaling information between transmitter and receiver, which can be included at MAC and higher layers. The signaling process should be undertaken in a capacity-efficient way. A promising research area is represented by distributed coordination mechanisms where some decision logic provides inherent coordination between transmitter and receiver.

The availability of inexpensive software radio platforms for a multiplicity of wireless applications will be one of the enablers of future multimedia communication systems.

Much work remains in developing good adaptive strategies. For multiuser systems adaptive modulation can be combined with other adaptive resource allocation policies like dynamic channel and base station assignment. Adaptive joint source and channel coding strategies that combine adaptive compression with adaptive modulation may also lead to good performance in time-varying channels. Furthermore, the cross-layer design is a methodology that requires further investigation, in order to lead to a scalable, robust and simple enough implementation.

Although the adaptive approach can be seen as a lower complexity alternative to diversity technique, it can be also applied to diversity techniques. Transmit diversity techniques that utilize CSI at the transmitter have shown considerable performance improvement over nonadaptive techniques [75, 76]. The adaptive antenna technology, which is one of the enabling technologies of future communication systems, has not been discussed in this chapter. An interested reader can refer to [77].

References

[1]   *IEEE Commun. Mag.*, "Feature Topic on Design Methodologies for Adaptive and Multimedia Networks," Vol. 39, No. 11, Nov. 2001, pp. 106–148.

[2]   Berezdivin, R., R. Breinig, and R. Topp, "Next Generation Wireless Communications Concepts and Technologies," *IEEE Commun. Mag.*, March 2002, pp. 108–116.

[3]   E. Cianca, et al., "Channel Adaptive Techniques in Wireless Communications: An Overview," *Journal of Wireless Communication and Mobile Computing*, Wiley, 2002.

[4]   Hanzo, L., C. H. Wong, and M. S. Yee, *Adaptive Wireless Transceivers: Turbo-Coded, Turbo-Equalised and Space-Time Coded TDMA, CDMA, MC-CDMA and OFDM Systems*, Wiley-Europe Publishers, Chichester, 2002.

[5]   Prasad, R., *Universal Wireless Personal Communications*, Norwood, MA: Artech House, 1998.

[6] Hayes, J. F., "Adaptive Feedback Communications," *IEEE Trans. Commun. Technol.*, Vol. COM-16, Feb. 1968, pp. 29–34.

[7] Cavers, J. K., "Variable-Rate Transmission for Rayleigh Fading Channels," *IEEE Trans. Comm.*, Vol. COM-20, Feb. 1972, pp. 15–22.

[8] Otsuki, S., S. Sampei, and N. Morinaga, "Square-QAM Adaptive Modulation/TDMA/TDD Systems Using Modulation Level Estimation with Walsh Function," *Electron. Lett.*, Vol. 31, Feb. 1995, pp. 169–171.

[9] Kamio, Y., et al., "Performance of Modulation-Level-Controlled Adaptive Modulation Under Limited Transmission Delay Time for Land Mobile Communications," *Proc. IEEE VTC'95*, July 1995, pp. 221–225.

[10] Webb, W. T., and R. Steele, "Variable Rate QAM for Mobile Radio," *IEEE Trans. on Comm.*, Vol. 43, July 1995, pp. 2223–2230.

[11] Vucetic, B., "An Adaptive Coding Scheme for Time Varying Channels," *IEEE Trans. on Comm.*, Vol. 39, May 1991, pp. 653–663.

[12] Alamouti, S. M., and S. Kallel, "Adaptive Trellis-Coded Multiple-Phased-Shift Keying for Rayleigh Fading Channels," *IEEE Trans. Comm.*, Vol. 42, June 1994, pp. 2305–2314.

[13] Ue, T., et al., "Symbol Rate and Modulation Level Controlled Adaptive Modulation/TDMA/TDD for High-Bit-Rate Wireless Data Transmission," *IEEE Trans. Veh. Tech.*, Vol. 47, No. 4, Nov. 1998, pp. 1134–1147.

[14] Goldsmith, A. J., and P. P. Varaiya, "Capacity of Fading Channels with Channel Side Information," *IEEE Trans. Inform. Theory*, Vol. 43, No. 6, Nov. 1997.

[15] Taricco, G., E. M. Biglieri, and G. Caire, "Impact of Channel-State Information on Coded Transmission over Fading Channels with Diversity Reception," *IEEE Trans. Commun.*, Vol. 47, No. 9, Sept. 1999, pp. 1284–1287.

[16] Caire, G., and S. Shamai, "On the Capacity of Some Channels with Channel State Information," *IEEE Trans. Inform. Theory*, Vol. 45, No. 6, Sept. 1999, pp. 2007–2019.

[17] Lee, C. Y., "Spectrum Efficiency in Cellular," *IEEE Trans. on Veh. Tech.*, Vol. 38, May 1989, pp. 69–75.

[18] Lee, H.-J., S. Komaki, and N. Morinaga, "Theoretical Analysis of the Capacity Controlled Digital Mobile System in the Presence of Interference and Thermal Noise," *IEICE Trans. Comm.*, Vol. E75-B, No. 6, June 1992, pp. 487–493.

[19] Torrance, J., and L. Hanzo, "Upper Bound Performance of Adaptive Modulation in a Slow Rayleigh Fading Channel," *Electronics Letters*, Vol. 32, April 11, 1996, pp. 718–719.

[20] Wong, C., and L. Hanzo, "Upper-Bound of a Wideband Burst-by-Burst Adaptive Modem," *Proceedings of VTC'99 (Spring)*, Houston, Texas, IEEE, May 16–20 1999, pp. 1851–1855.

[21] Wong, C., and L. Hanzo, "Upper-Bound Performance of a Wideband Burst-by-Burst Adaptive Modem," *IEEE Transactions on Communications*, Vol. 48, March 2000, pp. 367–369.

[22] Torrance, J., and L. Hanzo, "Optimisation of Switching Levels for Adaptive Modulation in a Slow Rayleigh Fading Channel," *Electronics Letters*, Vol. 32, June 20, 1996, pp. 1167–1169.

[23] Goldsmith, A. J., and S.-G. Chua, "Variable-Rate Variable Power MQAM for Fading Channels," *IEEE Trans. on Comm.*, Vol. 45, Oct. 1997, pp. 1218–1230.

[24] Torrance, J., and L. Hanzo, "Latency and Networking Aspects of Adaptive Modems over Slow Indoors Rayleigh Fading Channels," *IEEE Transactions on Vehicular Technology*, Vol. 48, No. 4, 1998, pp. 1237–1251.

[25] J. Torrance, L. Hanzo, and T. Keller, "Interference Aspects of Adaptive Modems over Slow Rayleigh Fading Channels," *IEEE Transactions on Vehicular Technology*, Vol. 48, Sept. 1999, pp. 1527–1545.

[26] Alouini, M., X. Tang, and A. J. Goldsmith, "An Adaptive Modulation Scheme for Simultaneous Voice and Data Transmission over Fading Channels," *IEEE Journal on Selec. Areas in Comm.*, Vol. 17, No. 5, May 1999, pp. 837–850.

[27] Calderbank, A. R., and N. Seshadri, "Multilevel Codes for Unequal Error Protection," *IEEE Transactions on Information Theory*, Vol. 39, No. 4, July 1993, pp. 1234–1248.

[28] J. K. Cavers, "An Analysis of Pilot Symbol Assisted Modulation for Rayleigh Fading Channels," *IEEE Trans. Veh. Technol.*, Nov. 1991, pp. 686–693.

[29] Andersen, J. B., et al., "Prediction of Future Fading Based on Past Measurements," *Proc. IEEE VTC'99*, Sept. 1999.

[30] Ekman, T., and G. Kubin, "Nonlinear Prediction of Mobile Radio Channels Measurements and MARS Model Design," *Proc. IEEE ICASSP'99*, pp. 2667–2670.

[31] Duel-Hallen, A., S. Hu, and H. Hallen, "Long Range Prediction of Fading Signals," *IEEE Signal Processing Magazine*, Vol. 17, May 2000, pp. 62–75.

[32] Lau, V. K. N., and S. V. Maric, "Variable Rate Adaptive Channel Trellis Coded QAM for High Bandwidth Efficiency Application in Rayleigh Fading Channels," *Proc. IEEE VTC'98*, pp. 348–352.

[33] Torrance, J., and L. Hanzo, "Demodulation Level Selection in Adaptive Modulation," *Electronics Letters*, Vol. 32, Sept. 12, 1996, pp. 1751–1752.

[34] Goldsmith, A. J., and S. Chua, "Adaptive Coded Modulation for Fading Channels," *IEEE Trans. On Comm.*, Vol. 46, No. 5, May 1998, pp. 595–602.

[35] Forney, G. D., Jr., et al., "Efficient Modulation for Band-Limited Channels," *IEEE J. Select. Areas Commun.*, Vol. SAC-2, Sept. 1984, pp. 632–647.

[36] Forney, G. D., "Coset Codes—Part I: Introduction and Geometrical Classification," *IEEE Trans. Inform. Theory*, Vol. 34, Sept. 1998, pp. 1123–151.

[37] Berrou, C., A. Glavieux, and P. Thitimajshima, "Near Shannon Limit Error-Correcting Coding and Decoding: Turbo Codes," *Proceedings of the International Conference on Communications*, Geneva, Switzerland, May 1993, pp. 1064–1070.

[38] Robertson, P., "Illuminating the Structure of Code and Decoder of Parallel Concatenated Recursive Systematic (Turbo) Codes," *IEEE Globecom*, 1994, pp. 1298–1303.

[39] Benedetto, S., and G. Montorsi, "Unveiling Turbo Codes: Some Results on Parallel Concatenated Coding Schemes," *IEEE Trans. Inform. Theory*, Vol. 42, March 1996, pp. 409–428.

[40] Divsalar, D., and F. Pollara, "Turbo Codes for PCS Applications," *Proc. 1995 IEEE Int. Conf. Communications*, Seattle, Washington, May 1995, pp. 54–59.

[41] Benedetto, S., and G. Montorsi, "Design of Parallel Concatenated Convolutional Codes," *IEEE Trans. Commun.*, Vol. 44, May 1996, pp. 591–560.

[42] Benedetto, S., R. Garello, and G. Montorsi, "A Search for Good Convolutional Codes to Be Used in the Construction of Turbo Codes," *IEEE Trans. Commun.*, Vol. 46, Sept. 1998, pp. 1101–1105.

[43] Lin, S., and D. J. Costello, *Error Control Coding: Fundamentals and Applications*, Englewood Cliffs, NJ: Prentice-Hall, 1983.

[44] Mandelbaum, D. M., "On Forward Error Correction with Adaptive Decoding," *IEEE Trans. Inform. Theory*, Vol. IT-21, March 1975, pp. 230–233.

[45] Haccoun, D., "High-Rate Punctured Convolutional Codes for Viterbi and Sequential Decoding," *IEEE Trans. Commun.*, Vol. 37, No. 11, Nov. 1989, pp. 1113–1125.

[46] Lau, V. K. N., "Channel Capacity and Error Exponents of Variable Rate Adaptive Channel Coding for Rayleigh Fading Channels," *IEEE Trans. Commun.*, Vol. 47, No. 9, Sept. 1999, pp. 1345–1356.

[47] Goeckel, D., "Adaptive Coding for Time-Varying Channels Using Outdated Fading Estimates," *IEEE Trans. Commun.*, Vol. 47, June 1999, pp. 844–855.

[48] Ungerboeck, G., "Channel Coding with Multilevel/Phase Signals," *IEEE Trans. Inform. Theory*, Vol. IT-28, Jan. 1982, pp. 55–67.

[49] Chase, D., "Code Combining—A Maximum-Likelihood Decoding Approach for Combining an Arbitrary Number of Noisy Packets," *IEEE Trans. Commun.*, Vol. COM-33, May 1985, pp. 385–393.

[50] Kallel, S., and D. Haccoun, "Generalized Type II Hybrid ARQ Scheme Using Punctured Convolutional Coding," *IEEE Trans. Commun.*, Vol. 38, Nov. 1990, pp. 1938–1946.

[51] Kallel, S., "Sequential Decoding with an Efficient Incremental Redundancy ARQ Scheme," *IEEE Trans. Commun.*, Vol. 40, Oct. 1992, pp. 1588–1593.

[52] Kallel, S., "Efficient Hybrid ARQ Protocols with Adaptive Forward Error Correction," *IEEE Trans. Commun.*, Vol. 42, Feb. 1994, pp. 281–289.

[53] Hagenauer, J., "Rate-Compatible Punctured Convolutional Codes (RCPC Codes) and Their Applications," *IEEE Trans. Commun.*, Vol. 36, Apr. 1988, pp. 389–400.

[54] Naryanan, K. R., and G. L. Stuber, "A Novel ARQ Technique Using the Turbo Coding Principle," *IEEE Commun. Letters*, Vol. 1, March 1997, pp. 49–51.

[55] Geraniotis, E., W. C. Chan, and V. D. Nguyen, "An Adaptive Hybrid FEC/ARQ Protocol Using Turbo Codes for Multi-Media Traffic," *IEEE ICUPC*, 1997, pp. 541–545.

[56] Kallel, S., "Complementary Punctured Convolutional Codes and Their Applications," *IEEE Trans. Commun.*, Vol. 43, No. 6, June 1995, pp. 2005–2009.

[57] Caire, G., and D. Tuninetti, "The Throughput of Hybrid-ARQ Protocols for Gaussian Collision Channel," *IEEE Trans. Inform. Theory*, Vol. 47, No. 5, July 2001, pp. 1971–1988.

[58] Leonardi, E., G. Caire, and E. Viterbo, "Modulation and Coding in Gaussian Collision Channels," *IEEE Trans. Inform. Theory*, Vol. 46, Sept. 2000, pp. 2007–2026.

[59] Babich, F., "Performance of Hybrid ARQ Schemes," *Proc. IEEE ICC 2001*, pp. 3036–3040.

[60] Chakraborty, S. S., and M. Liinaharja, "Performance Analysis of an Adaptive SR ARQ Scheme for Time-Varying Rayleigh Fading Channels," *Proc. IEEE ICC 2001*, Vol. 8, pp. 2478–2482.

[61] Kallel, S., S. Bakhtiyari, and R. Link, "An Adaptation Hybrid ARQ Scheme," *Wireless Personal Communications*, Vol. 12, No. 3, March 2000, pp. 297–311.

[62] Martins, J. A. C., and J. C. Alves, "ARQ Protocols with Adaptive Block Size Perform Better over a Wide Range of Bit Error Rate," *IEEE Trans. Comm.*, June 1990, pp. 737–739.

[63] Sastry, A. R. K., "Improving Automatic Repeat Request (ARQ) on Satellite Channels Under High Error Rate Conditions," *IEEE Trans. Commun.*, Vol. COM-23, 1975, pp. 436–439.

[64] Monecleay, M., and H. Bruneel, "Efficient ARQ Scheme for High Error Rate Channels," *Elettron. Lett.*, Vol. 20, 1984, pp. 986–987.

[65] Yao, Y.-D., "An Effective Go-Back.N ARQ Scheme for Variable Error Rate Channels," *IEEE Trans. Comm.*, May 1995, pp 20–23.

[66] Chakraborty, S., et al., "On the Performance of an Adaptive GBN Scheme in a Time Varying Channel," *IEEE Comm. Lett.*, April 2000, pp. 143–145.

[67] Annamalai, A., B. Bhargava, "Analysis and Optimisation of an Adaptive Multi-Copy Transmission ARQ Protocols for Time-Varying Channels," *IEEE Trans Comm.*, Oct. 1998, pp. 1356–1368.

[68] Rice, M., and S. B. Wicker, "Adaptive Error Control for Slowly Varying Channels," *IEEE Transactions on Communications*, Feb.–April 1994, pp. 917–926.

[69] Mitola, J., "The Software Radio Architecture," *IEEE Commun. Mag.*, May 1995, pp. 26–38.

[70] *IEEE Communications Mag.*, Special Issue on Software Radio, Vol. 37, No. 2, Feb. 1999, pp. 82–112.

[71] SDR Forum at http://www.sdrforum.org.

[72] Harada, H., and R. Prasad, *Simulation and Software Radio*, Norwood, MA: Artech House, 2002.

[73] Nola, K. E., et al., "Signal Space Based Adaptive Modulation for Software Radio," *Proc. IEEE WCNC2002 Conf.*, March 17–21, 2002, pp. 510–515.

[74] Rohani, K., M. Harrison, and K. Kuchi, "A Comparison of Base Station Transmit Diversity Methods for Third Generation Cellular Standards," *Proc. IEEE VTC'99*, pp. 351–355.

[75] Hu, S., et al., "Transmitter Antenna Diversity and Adaptive Signaling Using Long Prediction for Fast Fading DS-CDMA Mobile Radio Channel," *Proc. IEEE WCNC'99*, pp. 824–828.

[76] Alouini, M., and A. J. Goldsmith, "Capacity of Rayleigh Fading Channels Under Different Adaptive Tansmission and Diversity-Combining Techniques," *IEEE Trans. Veh. Technol.*, Vol. 48, No. 4, July 1999, pp. 1165–1181.

[77] Liberti J. C., and T. S. Rappaport, *Smart Antennas for Wireless Communications: IS-95 and Third Generation CDMA Applications*, Englewood Cliffs, NJ: Prentice-Hall, 1999.

6

Radio Resource Management for Wireless Multimedia Communications

6.1 Introduction

As evolution of wireless networks allows for an increasingly wide range of services, the Radio Resource Management (RRM) function, which divides the available resources amongst competing applications, is receiving increased attention. There are several reasons that render RRM very important [1–27]:

1. RRM functions allow the support of different requirements from the various services that the wireless network is required to support.
2. RRM may ensure the planned coverage (i.e., the area where the service is supported) for each service.
3. RRM may optimize capacity utilization.

Basically, the RRM has the complex task of maximizing the number of users that can be served satisfying their different service requirements, in time varying radio conditions and dynamic traffic behavior.

In this chapter we will focus on the main RRM functions and their implementation in different wireless networks.

The chapter is organized as follows. In Section 6.1 the general definition of the RRM problem is presented. Section 6.2 describes the most important RRM functions in GPRS. Important RRM issues in UMTS and in the future wireless systems are presented in Sections 6.3 and 6.4. Conclusions are drawn in Section 6.5. Before formulating the RRM optimization problem, basic concepts on QoS requirements are recalled in the next section.

6.1.1   QoS Requirements

In order to guarantee a satisfactory end user quality, the transmission of a data flow, which is originated by the application, has to satisfy certain requirements that define the QoS profile for the information data stream of interest. Usually, the QoS attributes for a particular application/service are: required throughput, maximum acceptable delay, maximum acceptable delay jitter, and maximum acceptable bit error rate.

From the end user point of view, the following issues have to be taken into account [5]:

- End users only care about the degree of QoS, and not about how it is provided.
- Only the QoS perceived by the end user matters.
- The number of "user-defined/user-controlled" parameters has to be a minimum.
- A derivation/definition of QoS attribute from the application requirements has to be simple.
- End-to-end QoS has to be provided.

A very frequent classification of the service class is based on the application/service delay requirement. For example, in the UMTS standardization the following four service classes are defined: *background class, interactive class, streaming class,* and *conversational class.* The main difference between these classes is the delay sensitiveness of the traffic. The conversational class is the most delay sensitive, while background class is the most delay tolerant. Table 6.1 summarizes the main characteristics of UMTS QoS classes.

In the standardization the definition of QoS attributes are the same for GPRS Release 99 and UMTS. The QoS attributes for GPRS Release 97/98 can be mapped on the Release 99 UMTS attributes as specified in [5]. However, the set of QoS attributes in UMTS is much larger than the set specified in GPRS Release 97/98 (Figure 6.1).

The throughput requirement is dependent on the information source. A throughput of 12 Kbps would be enough in order to transfer speech with GSM quality. For an audio stream with stereo quality, the throughput requirement is higher than 32 or 64 Kbps, while for a video-communication it is higher than 128 Kbps.

The bit error rate depends on the service class. For background (e-mail, FTP) and interactive (WWW browsing) services, the received data should be error-free. Due to the less stringent delay requirements of this service class, higher reliability can be achieved by error correction techniques (e.g., packet

Table 6.1
UMTS QoS Classes

Traffic Class	Conversational Class	Streaming Class	Interactive Class	Background Class
Fundamental characteristics	Preserve time relation (variation) between information entities of the stream	Preserve time relation (variation) between information entities of the stream	Request response pattern	Destination is not expecting the data within a certain time
			Preserve payload content	
	Conversational pattern (stringent and low delay)			Preserve payload content
Example of the application	Voice, video games, voice telephony	Streaming video	Web browsing, network games	Background download of e-mail

retransmissions). The use of error correction mechanisms is rather limited by the delay requirement (e.g., for the interactive class). For the streaming and conversational classes the acceptable BER depends on the type of information— namely, for speech, an acceptable BER is on the order of 10^{-2} or 10^{-3} and it may be even smaller for video transfer.

The application may specify its QoS requirements to the network by requesting a radio access bearer (RAB) with any of the specified traffic type, maximum transfer delay, delay variation, bit error rates, and data rates. In

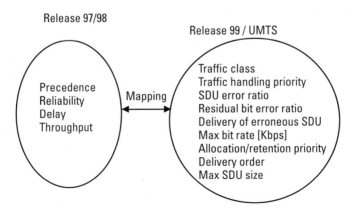

Figure 6.1 Mapping of QoS parameters.

practice, it should be possible to define the main RAB characteristics from the service quality requirements:

- The transmission rate of RAB should be determined by the bandwidth requirement of the information source.
- The choice of dedicated or shared RAB should be based on the requirement for the maximum delay and delay jitter.
- The SNIR requirement, channel coding, and interleaving for the RAB should be based on the BER requirement.

6.2   General Formulation of the RRM Problem

Let us assume a cellular network with M mobiles in the service area and denote with $B = \{1, 2, ..., B\}$ the set of all BSs used to provide the necessary coverage. Denote with C the number of available orthogonal channels in the system (i.e., the system capacity). The numbered set of all available channels is $C = \{1, 2, ..., C\}$. The channels orthogonality could be established in different ways, such as in time and frequency domain in GSM or in the code domain in WCDMA. In GSM, as a representative of 2G systems, there is an intrinsic upper limit on the system capacity since the upper bound is the number of frequencies multiplied with eight timeslots. On the other hand, in the WCDMA scheme, the set of orthogonal channels C is practically infinite and the capacity is determined by the interference condition in the system. The WCDMA system is, therefore, interference-limited.

The link (power) gain matrix G characterizes the radio conditions in the system:

$$G = \begin{pmatrix} G_{11} & G_{12} & \cdots & G_{1M} \\ G_{21} & G_{22} & \cdots & G_{2M} \\ \cdots & & & \\ G_{B1} & G_{B2} & \cdots & G_{BM} \end{pmatrix} \tag{6.1}$$

The matrix element G_{ij} represents the link gain between the BS i and MS j; M represents the number of active mobiles. The gain matrix G is dynamic, the dimension M is changing, based on the offered load, and each element G_{ij} changes with the mobile movement. The radio resource management, taking into account the link gain matrix G, assigns [3]:

1. One or more access points from the set B;
2. A channel from the set C;

3. The transmit power of the BS and of the mobile.

The assignments (1), (2), and (3) should maximize the number of users with a sufficient QoS. As outlined in the previous section, providing a stringent definition of the QoS for a communication service is a complex problem. A simple measure of the QoS, namely, the signal-to-interference + noise ratio (SNIR), is here considered. This measure is strongly connected with the performance measures as the bit or frame error probability. Therefore, assignments (1), (2), and (3) aim at maximizing the number of users for which the following inequality holds, for both the uplink (mobile-to-access port) and the downlink (access port-to-mobile):

$$SNIR_j = \frac{P_j G_{jj}}{\sum_{\substack{M \\ m \neq j}} P_m G_{jm} \theta_{jm} + N} \geq \gamma_j; j = 1, ..., M \tag{6.2}$$

In (6.2), $SNIR_j$ denotes the SNIR at the receiver; P_j is the transmitter power used by the end user j; θ_{jm} is the normalized cross-correlation between the signal of interest and the interfering signal from the mobile m (other than j); N denotes the thermal noise power at the access port, while γ_j is the target SNIR of the service that is being used by the mobile j.

6.3 Radio Resource Management in GPRS

Some of the new features of GPRS, with respect to GSM, are as follows [7]:

- The possibility of collecting physical channels (timeslots) for one user;
- The use of four different coding schemes (CS) for the radio transmission;
- The capability of sharing physical channels among several ongoing connections.

GPRS terminals, available on the market today with 4+1 multislot configuration and CS4 (i.e., 22.4 Kbps per timeslot), may achieve in the downlink rates of approximately 90 Kbps. The available radio resources in a GSM/GPRS network will be shared among the voice users and the GPRS data users (see Figure 6.2).

At the initial stage, GPRS data traffic is expected to be low and the blocking probability for voice users must not be affected. Therefore, the GPRS capacity should be allocated on a best-effort base. This means that the border that

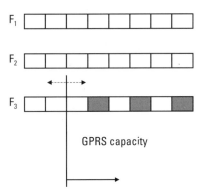

Figure 6.2 Capacity sharing between voice and GPRS users (GSM cell with three carrier frequencies).

defines the GPRS capacity should be adjusted according to the current number of voice users in the cell and the amount of GPRS traffic (see Figure 6.2). If the current voice traffic is very high, then the GPRS capacity might be reduced to zero.

However, with the development of wireless packet data services and the expansion of the GPRS customer population, wireless operators would like to introduce services with different QoS levels and guarantees for their customers. Therefore, the concept of RRM in GPRS for future applications may be explained in a more comprehensive manner through Figure 6.3.

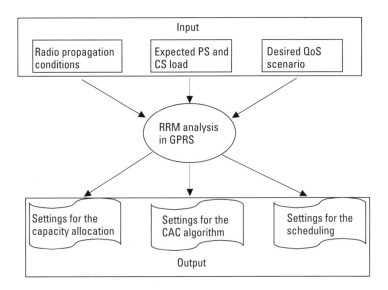

Figure 6.3 Concept of RRM in GPRS.

The GPRS network operator has the information about the current radio link quality, estimated level of packet switched and circuit switched load and the QoS requirements as an input to the process. Then, the RRM should provide the appropriate settings for the GPRS capacity, CAC, and scheduling, in order to efficiently support different services.

If only the best-effort traffic is supported by the GPRS system, then the CAC algorithm is not needed. All newly arrived calls will be accepted, and consequently, all ongoing calls will get a decreasing amount of capacity. CAC algorithm is unavoidable when GPRS operators want to support specific QoS levels. This algorithm is invoked whenever a new call arrives in the system (cell). It decides whether the call can be accepted or not, depending on the following two important factors (see Figure 6.4):

1. *Call QoS requirements:* type of the call (e.g., its priority level), delay property (e.g., conversational, interactive, streaming, or background), bandwidth requirements (e.g., average throughput).

2. *Cell current condition:* radio link quality (e.g., effective throughput per timeslot), current load conditions (e.g., number of ongoing GPRS data connections), and the total available capacity.

The CAC is a protection mechanism for the QoS of the ongoing calls in the system. In other words, this mechanism allows the network to reject calls before the QoS requirements of the ongoing calls and the QoS requirements of the newly arrived call cannot be longer satisfied.

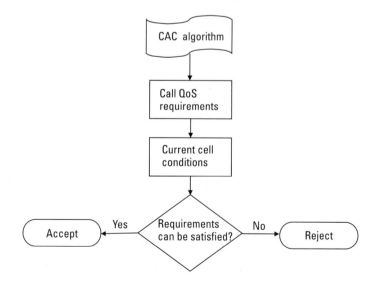

Figure 6.4 Schematic view of the CAC algorithm.

To support different delay requirements, the GPRS network may also use scheduling mechanisms, which give priority to the data streams of the services with more stringent delay requirements. Examples of scheduling algorithms with some degree of fairness are the round-robin scheduling policies. In combination with differentiated GPRS capacity allocation, which means allocation of physical channels in exclusive use for GPRS data streams, the more strict delay requirements of some application can be met. The trade-off is that this will heavily influence the blocking probability for voice users, as less physical channels will be available for voice communication.

Finding the appropriate setting for the GPRS capacity, CAC, and scheduling algorithms is not a straightforward task. First of all, proper models for the application traffic behavior are needed and they are usually complex. Performance assessment via analytical investigation, which takes into account these traffic models, the adaptive GPRS capacity allocation, and the specific CAC and scheduling policy, are difficult. Therefore, wireless operators may derive the appropriate thresholds for the RRM algorithms in GPRS through specific simulation studies [8].

6.4 RRM in UMTS

Wideband CDMA for the FDD mode and its hybrid associate time-division CDMA (TD-CDMA) for the TDD mode are key elements of the UMTS standard [9]. Since a CDMA-based system is interference-limited, there is a trade-off between the capacity and the quality supported by the system (i.e., the so-called "soft" capacity of UMTS). RRM algorithms aim at optimally using the soft CDMA capacity.

In order to better define the "soft" CDMA capacity [1], a capacity analysis in the uplink is considered. The CDMA system of interest has the following properties:

- Total system bandwidth W;
- One service with transmission rate R;
- μ is the target transmitter energy per bit versus interference ratio;
- p_{on} is the averaged activity factor for the service under investigation;
- Perfect UL power control, which means equal received power $P_{0,rx}$ at the BS from all mobiles belonging to the cell;
- Uniform traffic distribution with N users per cell.

More than 95% from the total interference comes from the six cells in the first tier and the 12 cells in the second tier around the reference cell, assuming regular hexagonal cellular structure. Therefore, the quality requirement in the uplink direction can be written as (not accounting for thermal noise):

$$\frac{E_b}{I_0} = \frac{W/R}{P_{oN}} \cdot \frac{P_{0,rx}}{\underbrace{(N-1)P_{0,rx}}_{same\ cell} + \underbrace{\sum_{j=1}^{6N} P_{j,rx}}_{1st\ tier} + \underbrace{\sum_{j=1}^{12N} P_{j,rx}}_{2nd\ tier}} \geq \mu^{UL} \tag{6.3}$$

where $P_{j,rx}$ is the interference power from user j received at the reference base station but power-controlled by its own base station. The total outer-cell interference is a sum of log-normally-distributed random variables. For large N, the outer-cell interference is usually modeled with Gaussian random variable with mean and standard deviation that depends on the number of users, radio channel characteristics, and service activity.

Denote with $P_{outage} = P_r (E_b/N_0 < \mu^{UL})$ the probability that a call has bad quality in UL. WCDMA capacity is usually defined as the maximum number of users N such that the P_{outage} is smaller than a predefined value (e.g., $P_{outage} < 0.05$). The soft WCDMA capacity comes from the drastic influence of the quality requirements on P_{outage} (and consequently on the capacity) and from the desired upper bound for P_{outage}. Figure 6.5 shows the P_{outage} for two different transmission rates (i.e., different processing gains, namely, 256 and 341.3 for

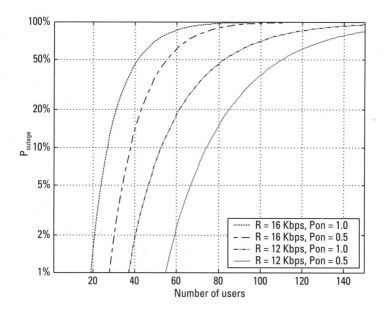

Figure 6.5 The influence of transmission rate, channel activity, and the number of users on P_{outage}.

R = 12 and R = 16 Kbps, respectively) and service activity factors, in case of Gaussian assumption for the total multiple access interference.

From (6.1) and Figure 6.5 we can conclude that the most important factors impacting the soft capacity of the CDMA system are the E_b / I_0 target, transmission rate (i.e., processing gain W/R), service activity, and the number of users in the system.

The main RRM functions in UMTS are power control, handover control, admission control, load control, and packet scheduler. They are located in different network elements and in the mobile terminal itself, as shown in Figure 6.6.

Since capacity in UMTS is interference-limited, transmission power levels of each radio connection should be kept as low as possible and yet high enough to satisfy the quality requirement (6.2). This is the task of the power control (PC) algorithm. In UMTS there are three loops of power control: open loop, outer loop, and inner loop. The open loop power control defines the initial transmission power level for UL (located in MS) and in DL (located in node B). The outer loop PC governs the quality target for the inner loop PC. In the UL it is executed in the RNC where the quality target, usually expressed with the energy-per-bit-to-interference density ratio E_b / I_0, is adjusted based on the frame erasure rate in UL. For the DL direction the outer loop is executed at the MS where the E_b / I_0 target for the DL is adjusted accordingly. The inner loop PC is executed each timeslot (with an update frequency of 1,500 Hz) for adjusting the transmission power of the MS and node B (i.e., base station node). Power control compensates for the fading characteristics of the radio channel for MS speeds below 80 km/h.

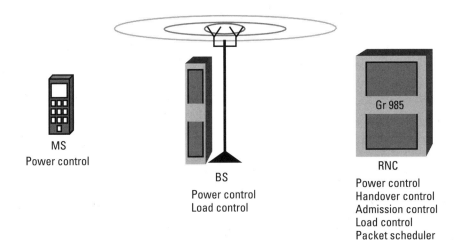

MS
Power control

BS
Power control
Load control

Gr 985

RNC
Power control
Handover control
Admission control
Load control
Packet scheduler

Figure 6.6 Typical locations for RRM functions in UMTS.

Multiplexing several services on a single physical connection is one of the requirements of UMTS. There will be only one inner loop PC for all these services so there is only one common target for the inner loop PC. This target is selected according to the service requiring the highest target (Figure 6.7).

The handover control in UMTS is performed to decrease the extensive transmission power in UL when the MS is moving at the edge of two or more cells. With soft handover (connections with different nodes B) and softer handover (connections with different sectors of one node B) the UL transmission is power controlled by the strongest node B and also the UL signals received at several nodes B (or sectors) can be further combined to improve the quality. The thresholds in the soft/softer handover algorithm determine when an MS is establishing or releasing a connection with additional nodes B (or sectors). The optimization criterion consists of minimizing the transmission power in UL and DL, which turns out in an increased system capacity. The so-called hard handover in UMTS, which may occur between two UMTS carrier frequencies, between the TDD and FDD mode, and between the UMTS[1] and GSM/GPRS systems, could be based on traffic load conditions. In fact, wireless operators might be interested in having traffic load sharing among different wireless systems and optimizing the usage of the available radio resources.

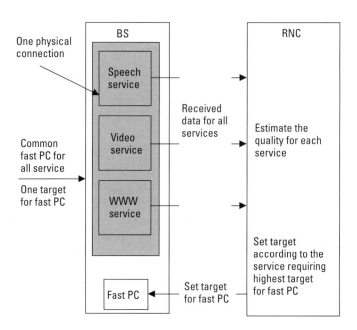

Figure 6.7 UL outer power control for multiple services on one physical connection.

---

1. In the future, intersystem handovers towards WLAN or HIPERLAN/2 can be expected.

CAC has the preventive role of keeping the UMTS network from entering an overload state, thus protecting the QoS of ongoing connections. If such an overload situation still occurs, however, then the load control function takes care that the UMTS network is brought back to a stable operating mode. The load control functions are taking the following actions:

1. Node B may deny sending power-up commands in DL and ignore the received power-up commands in UL.
2. RNC may lower the SNIR targets for the inner loop PC in UL, lower the bit rates for real-time users, reduce the DL packet data transfer for non-real-time services via the scheduling algorithm, perform UMTS interfrequency or intersystem handover to GSM, and, as a last measure, start dropping users in a controlled manner.

The CAC algorithm in UMTS has the standard functionality as presented in Figure 6.4; that is, to estimate the capacity requirement for the newly arrived (or handover) call, to measure continuously the load in UL and DL, and to decide if the admitted request can be granted or rejected.

The interactions of the different RRM algorithms that consider the link gain matrix G, available channel set C and the desired QoS for particular service are presented in Figure 6.8.

Three steps are identified:

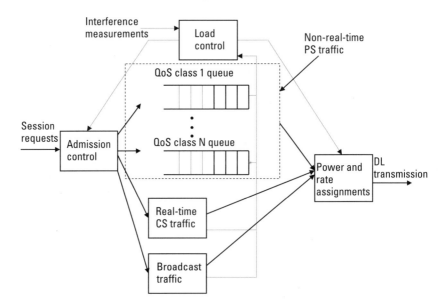

Figure 6.8 Interaction of RRM algorithms in UMTS.

1. At the session arrival, the CAC decides if and where (at which node B) the new (or handed over) session is accepted or rejected. The CAC is controlled by the load control function.

2. Assignment of the appropriate amount and type of radio channels from the set *C.* The very important issue here is to choose the most appropriate transport channel for the service (i.e., shared channels for non-real-time PS data, dedicated channels for real-time connections, or broadcast transport channels for multimedia broadcast messaging). The load measurements are sent to the load control functions.

3. Allocation of the DL transmission power and rate, which is also monitored by the load control.

As shown in Figure 6.8, after admitting the session request in the system, the RRM logic has to allocate the appropriate amount and type of wireless resources. This is very important in UMTS since it will support a wide range of services. Therefore, in the UMTS standard, the concept of logical, transport, and physical channels is introduced. Figures 6.9 and 6.10 give a general overview of the logical, transport, and physical channels in UL and DL, respectively. Some introductory concepts on UMTS and in particular on the RRM function in UMTS have been already discussed in Chapters 1 and 2.

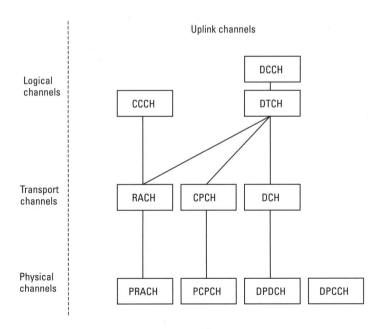

Figure 6.9 Uplink logical, transport, and physical channels.

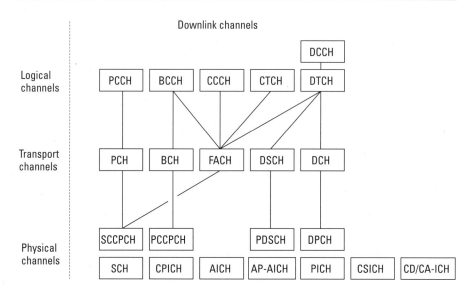

Figure 6.10 Downlink logical, transport, and physical channels.

Note that in both UL and DL the mapping of transport channels onto physical channels is unique [10]. The logical channel describes the logical context of the data that is transferred with that channel. Logical channels are divided in dedicated (destined for single user) and common (destined for group of users) channels. Furthermore, dedicated and common logical channels may transfer data traffic or control information from the higher layers.

The transport channels describe the way in which the data is transferred over the air. In the UL there are the following transport channels: random access channel (RACH), common packet channel (CPCH), and dedicated channel (DCH). In the DL direction there are the paging channel (PCH), broadcast channel (BCH), forward access channel (FACH), downlink shared channel (DSCH), and the DCH.

Note that some of the transport channels, such as PCH and BCH, are used to carry only relevant UMTS system information and not end user data. Furthermore, some of the physical channels in Figures 6.9 and 6.10 do not have associated transport channels. They are introduced for proper functioning of the UMTS system. For example, the synchronization channel (SCH) in the physical layer is used to synch the channelization code at the MS receiver with the incoming signal and to be able to despread the received signal properly. The physical control channel (PDCH) conveys the PC command bits, the pilot bits used for radio channel (SNIR) estimation and the transport format of the transmitted radio block. The common pilot channel (CPICH) is used to initially camp on a UMTS cell and also in the handover process and so forth.

For transmitting data with multimedia content over the UMTS radio interface, the transport channels RACH, CPCH, FACH, DSCH, and DCH are of special interest. The FACH requires extra attention since several types of logical channels can be mapped onto it (Figure 6.10). Table 6.2 shows the attributes of all these transport channels. There should be an optimal mapping of application/service data and the corresponding UMTS transport channel. This can be done by extensive simulation investigation and validation with test beds. Each of these transport channels has different radio performance in terms of achievable BER, transmission rates, delays, and so forth. Moreover, the utilization of these channels should be considered in finding the optimal mapping. This is an interesting field for RRM investigation.

Many services have dynamic (bursty) behavior and also the radio conditions in UMTS are very volatile depending on the propagation and traffic load conditions in the reference cell and the surrounding ones. Therefore, it could be advantageous to have RRM monitoring and performing transport channel switching during the lifetime of a call.

The dynamic transport channel switching is another interesting research area in the RRM field. The reasons for changing the transport channel type are to maintain the desired QoS of the connection and to improve the radio resource utilization. Considering the properties of the transport channels (see Table 6.2) and the application/service requirements, the transport channel switching can be influenced by the following factors:

Table 6.2
Properties of the UMTS (FDD) Transport Channels

Property	DCH	RACH	FACH	CPCH	DSCH
UE state	Cell_DCH	Cell_FACH	Cell_FACH	Cell_FACH	Cell_FACH
Direction	UL and DL	UL	DL	UL	DL
Power control	Inner loop	Open loop	Open loop	Inner loop	Inner loop
SHO	Yes	No	No	No	No
Applicability for bursty traffic	Poor	Good	Good	Good	Good
Target data volume	Medium/high	Small	Small	Small/medium	Medium/high
Setup time	High	Low	Low	Low	Low
Relative radio performance	High	Low	Low	Medium	Medium/high

1. The trade-off between the gain of a transport channel switching and the introduced overhead (signaling and switching delay).

2. Services with stringent delay and delay jitter requirements should use DCHs. This is valid for conversational and also streaming type of services. Services belonging to these two traffic classes are not suitable for transport type channel switching since the setting up and releasing of DCH takes a considerable amount of time, if compared with the maximum delay and delay jitter requirements. Furthermore, radio performance is crucial for this type of application, which can benefit from inner loop PC and soft/softer handover by using DCHs. As the service requirement on the maximum tolerable delay and jitter becomes less stringent, the service becomes more suitable for the shared/common transport channels.

3. For interactive and background type of services, common transport channels, such as RACH in UL and FACH in DL, and shared transport channels, such as CPCH in UL and DSCH in DL, are more suitable from the radio resource utilization point of view. For example, with this mapping the channelization codes in DL are more efficiently used and less interference is generated when compared to the use of DCHs. This would result also in higher capacity. However, which transport channel is used and switching to DCH is also influenced by the following factors:

- For relatively small amounts of data, the cost of setting up a DCH may be significant, and also, the benefit of inner loop PC is not that high, since during the short transmission time the radio conditions are almost constant and an open loop PC is sufficient. In this case, the choice of RACH and FACH would be appropriate.

- For relatively large amounts of data, the advantages of fast power control may be the decisive factor for choosing CPCH and DSCH.

- For interactive types of services, switching to DCH could be more appropriate in case of high load conditions on common or shared channels and in case of bad radio conditions at the cell edge (so the benefit of soft/softer handover could be utilized).

It is worth mentioning that in 2002, 3GPP was actively working to standardize the High-Speed Downlink Packet Access (HSDPA), which in turn introduces a new transport channel high speed downlink shared channel (HS-DSCH). The main goal of HSDPA is to support downlink peak rates in the range of 8 to 10 Mbps for best-effort packet data services. The higher data rates are achieved with the following physical layer enhancements [11]:

- Higher-order modulation (8-PSK, 16-QAM, and 64-QAM) combined with fast link adaptation (i.e., adaptive coding and modulation based on actual radio channel conditions) and fast hybrid ARQ (soft combining retransmitted packets to improve the BLER performance).
- HS-DSCH will not use inner loop (fast) downlink TCP, thus it will use in principle constant transmission power.
- Scheduling function (and the hybrid ARQ) located in node B and not at the RNC with short transmission time interval (TTI around 2 ms).
- Fast cell selection based on CPICH measurements to select at each radio frame transmission toward the MS, the node B with the best radio conditions.

An interesting RRM problem for the downlink shared transport channels DSCH and HS-DSCH is the choice of the scheduling scheme. This is important because if there is a high use of Internet-based applications, the amount of traffic in the DL direction could be much higher than in the UL direction. The scheduling of this traffic is a trade-off between the fairness among users transmitting on the shared transport channel in the DL and the need of maximizing its total DL throughput [12]. The use of standard round robin (RR) scheduling discipline is suitable to achieve fairness in the scheduling policy. Furthermore, if different QoS classes are using the shared transport channel, the appropriate prioritization between the different classes can be achieved by assigning weights in the RR scheduling. As the assignments for DL transmission with RR scheduling is independent of the radio channel condition, the total DL throughput might be decreased for a MS that experiences bad radio conditions. Therefore, a scheduling discipline based on the radio channel quality may optimize the total DL throughput of the transport channel. This scheduler allows DL transmission to the MS that has (instantaneously) the best radio channel conditions. The drawback of this policy is that it introduces high variability in the experienced throughput since the users with low radio channel quality will have significantly degraded quality. The compromise between fairness and consideration of the radio channel quality is the so-called proportional fair policy introduced in [12]. The results in [12] compare the performance of these policies and also the system capacity and percentage of satisfied users with different upper limits for the DL transmission power of the HS-DSCH. Note that the scheduler based on radio channel conditions is more suitable for HS-DSCH, since DSCH has not up to date information about these conditions.

6.5   RRM in Future Wireless Systems

In this section, future RRM developments in wireless LANs and general RRM issues in mobile ad hoc networks (MANETs) are presented.

The Broadband Radio Access Network (BRAN) working group in ETSI is standardizing a wireless LAN for broadband radio access up to 54 Mbps. This new standard is called HIPERLAN/2 and includes physical layer, radio link control, and data link control standards (see [13–17]). The interfaces toward other networks (e.g., UMTS) are made via specific design of convergence sublayers. The physical layer of HIPERLAN/2 is aligned with the physical layer of IEEE 802.11a wireless LAN system [17]. Note here that wireless broadband networks based on the IEEE 802.11a standard became commercially available in 2002. Among the most important RRM functions in these WLAN systems are the link adaptation function and the radio resources allocation in the MAC frame.

In HIPERLAN/2 and IEEE 802.11 systems there are different pairs of modulation and coding schemes possible. Each pair results in different transmission rate and packet error rate (PER) performance depending on the radio channel quality (i.e., SIR). The link adaptation scheme can dynamically change the pair modulation/coding scheme to optimize the throughput based on measured PER, signal level, and packet size. Extensive analysis of the physical layer performance of these two WLAN systems can be found in [15].

The MAC layer at the two WLAN standards is different. HIPERLAN/2 uses TDMA/TDD medium access with the frame structure as presented in Figure 6.11.

The allocation of radio resources is centrally scheduled by the AP, which allows for implementation of scheduling algorithms, QoS differentiation, and resource reservation for services with stringent delay and delay variation requirements.

The duration of the BCH is fixed. Through these channels the relevant system information is conveyed to the terminals. The duration of the FCH, DL and UL phase, direct link (DiL) phase, and RCH is dynamically adapted to the current load conditions of the access point. DiL phase is present if there are

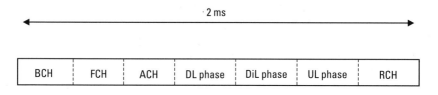

Figure 6.11   HIPERLAN/2 MAC frame structure.

mobile terminals directly in communication on a peer-to-peer basis. Requests for the radio resources are signaled via the RCH, where contention for timeslots is present. If scheduled data is transmitted, then FCH is present and signals the frame structure. The transmission in the DL phase (from access point to the terminals) and UL phase (from terminals to the access point) is contention-free. The allocation of resources is signaled via the access feedback channel, as an answer to the resource request received via the RCH from the previous MAC frame.

The IEEE 802.11a standard, however, has a MAC scheme based on CSMA/CA. A mobile terminal before transmission of data senses the radio channel. If the channel is free then the transmission can commence. Otherwise, an exponential back-off period is implemented before attempting the following packet transmission. This type of MAC (also known as distributed coordination function) makes the IEEE 802.11a standard more suitable for ad hoc wireless networks and non-real-time (background or interactive) type of applications (see Chapter 2). It should be mentioned that the standard also has contention-free MAC via the point coordination function (PCF), but this alternative is optional even though it could support real-time services. The advantage of the CSMA/CA is, however, the avoidance of centralized scheduler that coordinates the radio transmissions.

MANETs are also receiving attention recently. In IETF there is a special working group (MANET working group) that investigates the routing protocols in ad hoc networks. These networks have been described:

A "mobile ad hoc network" (MANET) is an autonomous system of mobile routers (and associated hosts) connected by wireless links—the union of which form an arbitrary graph. The routers are free to move randomly and organize themselves arbitrarily; thus, the network's wireless topology may change rapidly and unpredictably. Such a network may operate in a standalone fashion, or may be connected to the larger Internet. (*Source:* IETF MANET group.)

Currently, there are many research activities in RRM field for MANETs. Due to the specific characteristics of MANETs such as dynamic topology, limited node performance, distributed algorithms, and so forth, the development of RRM functions is a difficult task. The most important factors in MANETs are the coverage (or connectivity) and the capacity [18, 19]. An extensive field of research is the QoS aware routing (see [20] for one typical example) and MAC with QoS support [21]. A common MAC design for MANETs is driven by the hidden/exposed terminal problem. The QoS constraints for particular applications require from the MAC layer additional functionalities to provide certain guarantees and to make distinctions among different types of connections.

Furthermore, MAC has to interact with QoS-aware routing in an appropriate way for providing the required communication quality over the whole path from source to destination (i.e., over the multiple hops). The current MAC proposals [21] that support QoS provisioning and differentiation are based on resource reservation along the connection path, and differentiation between non-real-time and real-time connection establishment and maintenance.

The recent important development in the RRM field for multimedia wireless communication is for systems beyond UMTS. In these systems the last wireless hop towards the end user could be carried over different radio access networks. Here, the interworking of UMTS, WLAN or HIPERLAN/2, and GSM/GPRS networks will play an important role. For example, the wireless operators could have in their coverage areas multiple possibilities for wireless communications via different radio access networks as presented in Figure 6.12.

In this type of wireless network, the setup of the common RRM (CRRM) functions will play an important role in the efficient utilization of the available radio bandwidths. To achieve this goal, the CRRM will perform traffic addressing towards less loaded wireless access systems. CRRM will choose the right radio access technology based on service requirements, current wireless system load, propagation conditions, interference, and capacity cost induced in the wireless network. This field is very challenging and interesting for future RRM research.

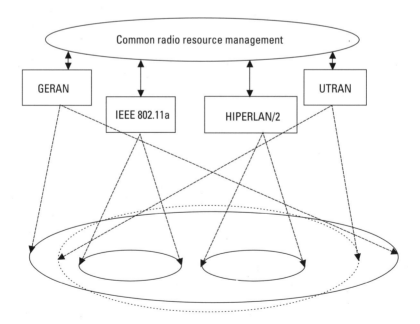

Figure 6.12 Interworking of different wireless access networks and CRRM.

6.6 Conclusions

This chapter has presented the most important RRM algorithms in the GPRS, UMTS, and future wireless communication systems. The first very important RRM algorithm for supporting multimedia communication is CAC. The CAC is crucial for protecting the QoS of ongoing connections and optimal usage of radio resources. For a proper operation of CAC, it is crucial that the network can estimate as accurately as possible the consumption of radio resources of a service and the current state of the wireless system in terms of traffic load, interference conditions, and capacity costs. These basic requirements for CAC are valid for all wireless systems. The wireless network planners will have to design the appropriate thresholds for the CAC decision logic in order to have satisfactory radio resource utilization and at the same time the desired grade of service and coverage.

Other important functions of the RRM are traffic scheduling, transport channel allocation and switching, handover control, and link adaptation. The trade-off that characterizes the RRM is between satisfying the QoS requirement and optimal usage of the radio resources.

GPRS has limited multimedia support due to the relatively low transmission rate of the radio link, and thus the requirements for RRM are minimal. On the other hand, RRM mechanisms are very important in UMTS and beyond. Further research is needed to find the optimal radio resource allocation procedures in UMTS, WLANs, and heterogeneous wireless coverage. For example, finding the optimal mapping between multimedia application and transport channels in UMTS, based on the achieved end-to-end QoS and resource utilization, is an issue that requires further investigation. Radio resource allocation in the MAC layer for WLANs and also MAC design complemented with QoS-based routing in ad hoc wireless networks are also promising topics for future research.

Finally, the role of common radio resource management is foreseen as crucial in heterogeneous wireless networks. Here, issues such as traffic addressing, handover control, allocation of wireless link over the most optimal radio access network based on capacity consumption, and resulting cost for the end user represents interesting fields for further research.

References

[1]   Jorguseski, L., J. Farserotu, and R. Prasad, "Radio Resource Allocation in 3rd Generation Mobile Communication Systems," *IEEE Commun. Mag.*, Vol. 39, No.2, Feb. 2001.

[2]   Zander, J., "Radio Resource Management—An Overview," *IEEE VTC '96*, Vol. 1, pp.16–20, May 1996.

[3]   Zander, J., "Radio Resource Management in Future Wireless Networks: Requirements and Limitations," *IEEE Commun. Mag.*, Aug. 1997, pp. 30–36.

[4]   Zander, J., "Radio Resource Management in 3rd Generation Personal Communication Systems," *IEEE Commun. Mag.*, No. 8, Aug. 1998.

[5]   3GPP Technical Specification Group, "QoS Concept and Architecture," (3GTS23.107), Oct. 1999.

[6]   3GPP Specification TS 23.107, "UMTS QoS Concept and Architecture," R'99/R4.

[7]   GSM 02.60, "GPRS Service Description—Stage 1," Version 7.5.0, Release 1998.

[8]   Stuckmann, P., and O. Paul, "Dimensioning GSM/GRPS Networks for Circuit-Switched and Packet-Switched Services," *Proc. WPMC '01*, Aalborg, Denmark, Sept. 2001.

[9]   Holma, H., and A. Toskala, *WCDMA for UMTS—Radio Access for Third Generation Mobile Communications*, New York: John Wiley & Sons, 2001.

[10] 3GPP TS 25.211, "Physical Channels and Mapping of Transport Channels onto Physical Channels (FDD)," v.5.1.0, June 2002.

[11] GPP TR 25.848, "Physical Layer Aspects of UTRA High Speed Downlink Packet Access," v4.0.0, March 2001.

[12] Parkvall, S., etal., "Evolving WCDMA for Improved High Speed Mobile Internet," *Future Telecommunication Conference*, Beijing, China, Nov. 28–30, 2001.

[13] ETSI TS 101 475, "Broadband Radio Access Network (BRAN); HIPERLAN Type 2; Physical (PHY) Layer," v1.1.1, April 2000.

[14] ETSI TS 101 761-1, "Broadband Radio Access Network (BRAN); HIPERLAN Type 2; Data Link Control (DLC) Layer; Part1: Basic Data Transport Functions," v1.1.1, April 2000.

[15] Doufexi, A., et al., "A Comparison of the HIPERLAN/2 and IEEE 802.11a Wireless LAN Standards," *IEEE Communications Magazine*, May 2002.

[16] ETSI TS 101 761-2, "Broadband Radio Access Network (BRAN); HIPERLAN Type 2; Data Link Control (DLC) Layer; Part 2: Radio Link Control (RLC) Sublayer," v1.1.1, April 2000.

[17] IEEE Std 802.11 a/D7.0-1999, "Part 11: Wireless LAN Medium Access Control (MAC) and Physical Layer (PHY) Specifications: High Speed Physical Layer in the GHz Band."

[18] Gupta, P., and P. R. Kumar, "The Capacity of Wireless Networks," *IEEE Transactions on Information Theory*, Vol. 46, No. 2, March 2000, pp. 388–404.

[19] Zahedi, A., and K. Pahlavan, "Capacity of a Wireless LAN with Voice and Data Services," *IEEE Transactions on Communications*, Vol. 48, No. 7, July 2000.

[20] Sinha, P., R. Sivakumar, and V. Bharghavan, "CEDAR: A Core-Extraction Distributed Ad Hoc Routing Algorithm," *IEEE Infocom '99*, New York, March 1999.

[21] Lin, C. R., and M. Gerla, "MACA/PR: An Asynchronous Multimedia Multi-Hop, Wireless Network," *Proceedings of IEEE INFOCOM '97*, 1997.

[22] Berg, M. "A Concept for Hybrid Random/Dynamic Radio Resource Management," *Proc. IEEE PIMRC'98*, Boston, 1998.

[23] Mihailescu, C., X. Lagrange, and P. Godlewski, "Dynamic Resource Allocation in Locally Centralized Cellular Systems," *Proc. VTC'98*, Ottawa, Canada, 1998, pp. 1695–1700.

[24] Katzela, M. N., "Channel Assignment Schemes for Cellular Mobile Telecommunication Systems: Comprehensive Survey," *IEEE Personal Communications*, Vol. 3, No. 3, June 1996.

[25] Zander, J., "Performance Bounds for Joint Power Control and Link Adaptation for NRT Bearers in Centralized (Bunched) Wireless Networks," in *Proc. PIMRC'99*, Sept. 1999.

[26] Zander, J., "Trends in Resource Management Future Wireless Networks," *IEEE Wireless Communications and Networking Conference, WCNC. 2000*, Vol. 1, pp. 159–163.

[27] Zander, J., S.-L. Kim, and M. Almgren, *Radio Resource Management for Wireless Networks*, Norwood, MA: Artech House, 2001.

7

Real-Time Services

7.1 Introduction

Real-time services involve the transmission of information, typically voice or video, which is intolerant of delay. With the scarcity of radio spectrum, one of the most important issues for real-time services, such as high-quality video and audio information, is to reduce the amount of data needed for representing the information without harming the subjective quality appreciably. Efficient video coding algorithms have been developed, which can compress video to a few kilobits per second with acceptable quality for many applications [1]. However, the penetration of wireless networks and the migration of the core network towards a packet-based transmission (i.e., IP, ATM) pose new challenges at all levels in the system design for real-time services [2–4]. First of all, there is a need for new low bit rate video coding techniques that are much more robust to channel degradation, to provide acceptable quality video even under adverse channel conditions [5–7]. Two further key requirements for future audio/video applications running over the Internet are *interactivity* and *scalability*. Interactivity provides the application with the capability to respond to any action triggered by the end user; the feature of scalability offers the possibility to modify a component of the video/audio information to fit the available resources. For instance, the size of an image can be changed in terms of number of pixels per width or height; in case of an image sequence, the quality can be scaled by altering the number of pictures transmitted or displayed per second. Video coding standards previous to MPEG-4 can hardly provide interactivity and have very poor scalability properties.

Furthermore, packet networks originally designed for non-real-time data services are not well suited for supporting real-time services. The best-effort approach used in the Internet cannot provide QoS guarantees to these delay-sensitive applications. Furthermore, the design of channel source/coding techniques changes when accounting for transport-delay constraints and considering that *not all the bits are equal*, and hence, some information should be unequally protected/compressed. Therefore, new signal processing techniques and new communication protocols are needed.

The structure of this chapter is as follows. The main differences in the design of packet networks for real-time services as compared to the current design for non-real-time services are highlighted in Section 7.2. Key concepts, which are extensively developed in this chapter, such as multiresolution decomposition and unequal error protection (UEP) are introduced. The basics of video compression are provided in Section 7.3. Standards, protocols, and technologies needed to support new video applications over the Internet are introduced in Section 7.4. Finally, the main technologies to support voice over IP are presented in Section 7.5. Open issues are drawn in Section 7.6.

7.2 Packet Networks for Real-Time Services

Real-time applications, like audio and video, are currently mostly transported across circuit-switched systems. This provides dedicated bandwidth per data stream—allocated for at least as long as the delivery lasted—without any bandwidth competition with other traffic.

The explosive growth of the Internet is changing this picture. Increasingly, interactive applications are being operated over the Internet and over private IP-based networks, such as desktop conferencing, videophone, Internet video (i.e., Web pages with embedded video). From an economical point of view, this use of packet networks provides several advantages:

- The communication service is billed at a flat monthly rate or connect time, not on the amount of information sent or the location of the called party.

- The user has full control over what is to be received and when.

- The choice of content is larger than by any other current delivery means.

Traditional packet-switched networks, however, are designed to work satisfactorily for bursty data services. Real-time services have quite different features and requirements and several changes at different levels are needed to adapt packet networks to real-time services.

In this section the main differences in the design of a packet network for real-time and non-real-time services are highlighted. However, one of the main challenges of future multimedia systems is to support efficiently both types of services.

Compared with circuit-switched networks, packet networks introduce delay jitter and possible congestion loss, to which data services are usually tolerant. For real-time services the delay jitter is usually removed by buffering the data before the audio and video reconstruction. However, there is an upper bound on the delay jitter that can be tolerated, which depends on the amount of buffering that the users will tolerate before the service becomes unusable.

Furthermore, transport-delay bounds are required, and packets arriving after a prescribed delay bound are usually considered fully lost. To achieve a given reliability in the presence of congestion loss and packet corruption, retransmissions or powerful error correcting codes are needed. However, error control techniques may introduce excessive delay. The choice of error control coding strategies, like forward error correction versus ARQ or more powerful hybrid FEC/ARQ (see Chapters 4 and 5), is influenced by these transport-delay requirements, which depend on the application. For example, video-on-demand is quite tolerant to delay; multimedia editing or video conferencing application requires transport-delay less than around 50 ms.

Packet-switched systems can introduce [8]:

- *Packet loss:* when the packet does not arrive because of buffer overflow or bit errors in the header;
- *Packet corruption:* when bit errors occur within the information data.

In data services any packet is equally important. Therefore, packet networks that are designed for data services do not distinguish between packet loss and packet corruption, as the packet that is corrupted is usually discarded. As a consequence, any lost or corrupted packet must be recovered by retransmission.

Real-time services, like audio and video, can tolerate some level of loss without excessive subjective impairment and should handle the effects in different ways. In fact, displaying the corrupted data as if it were correct, instead of discarding and masking them, may result in less severe subjective impairment than if the corrupted data were discarded and masked. On the other hand, lost data must be masked, for example in video by repeating information from a previous frame or in audio by substituting a zero-level signal.

An important property of voice and audio services is that *all bits are not equal*. For example, in the case of speech, the most important part of the signal (i.e., the part can be mainly distinguished by the human ear) is carried by some frequencies only. This property can be usefully exploited by techniques such as multiresolution decomposition and unequal error protection, which allow for separating signal components by significance, applying a different degree of protection on each component. Multiresolution decomposition and UEP, which are key concepts in all the compression techniques developed to deliver video and audio, are introduced in the following sections.

7.2.1 Multiresolution Decomposition

Multiresolution decomposition is a technique of successive approximations to a signal [9]. An ordered set of subsignals may be derived such that the first is a coarse approximation, which contains the basic information needed to reconstruct the source (e.g., motion vectors and low-frequency information), and by adding subsequent components (e.g., high-frequency residue details), the quality is increased with a decreasing marginal improvement. A multiresolution approach is well suited in many applications. For example, while browsing through image databases in the Web, only a coarse version of the full image can be downloaded, thus saving a noticeable amount of time. Then, one can fetch the rest of the image, or additional details, if the image seems of interest.

A multiresolution approach can be applied to source coding, thus enabling efficient adaptation to changing source bit rates and available bandwidths. For example, the higher-order components or coefficients of the decomposed source signal have lower variance and can be compressed efficiently by entropy coding (e.g., Huffman). The multiresolution decomposition also applies to channel codes, thus providing UEP.

7.2.2 Unequal Error Protection

UEP codes are families of channel codes that can change their coding strengths flexibly and efficiently. Examples of these codes are the rate compatible punctured convolutional (RCPC) codes [10]. Such code families can provide a hierarchy of *resolutions* of noise immunity, thus enabling an efficient adaptation to varying channel conditions. For instance, they are used in type II hybrid ARQ schemes to provide incremental redundancy (see Section 5.3.2). However, they can also be used to unequally protect different source components of the same signal having different error sensitivities. In fact, the higher-order components of a multiresolution decomposition contribute less to the signal in a mean squared error sense. Therefore, a more efficient channel coding may be designed, which concentrates losses on these subchannels and uses a powerful error correction code on the high-priority components [11]. This approach naturally leads to the concept of joint-source/channel coding [12], which has been used in packet video and is called hierarchical, embedded, or layered coding [13, 14].

While RCPC codes are powerful, they are not optimally suited when a code that is simultaneously decodable at multiple resolutions is needed.

Examples of such a scenario are broadcast and multicast, where users have different channel capacity. In a broadcast scenario a low-rate code of the RCPC family has to be used in order to protect the information of the worst channel. The redundancy introduced by this low-rate code is unnecessary for users with better channel conditions, thus leading to a waste of bandwidth. In general, in all the scenarios where it would be more efficient if the receiver decides the level of overhead that has been introduced in the received data stream, *embedded codes* are an attractive solution. A two-level embedded unequal error protection code can be described as an (n, k_1, k_2, t_1, t_2) code. The codeword length of the data stream is n; t_i represents the number of channel errors that the code with code rate k_i/n can correct. For instance, if t_1 t_2, the user with better channel conditions decodes the received data stream using the code rate k_1/n, while the users with worse channel conditions decode the same data stream using the lower rate code k_2/n. Embedded codes can be also useful for providing adaptability to channel conditions in point-to-point communications where there is no feedback channel information. However, embedded UEP codes are difficult to find and no structured method has been conceived to design them.

7.3 Video Compression Techniques

Video compression techniques aim at reducing the amount of data required to represent the video signal. A color TV quality image typically contains about 720 × 480 pixels, each pixel requiring 16 bits of resolution (8 for luminance and 8 for chrominance), giving a total of 5.5 Mbits of information. For broadcast quality TV, to portray full motion video, images must be presented at a frame rate of 30 frames per second. Hence, TV resolution digital video contains 166 Mbits of information per second. A 10-second clip of raw video hence requires 201 MB of storage space. The need to reduce the amount of data required to represent the video signal, without harming the subjective quality appreciably, clearly results. This objective can be accomplished by removing signal components that are subjectively unimportant and by exploiting the spatial and temporal correlation between video frames in order to remove the predictable or redundant signal content and transmit only the unpredictable information.

In this section, well-established and emerging concepts on video compression are presented. We will mainly refer to block-based coders, which are by far the most popular of the current video coding techniques. In a block-based coder, each frame is divided into rectangular units of 16 × 16 pixels, known as a *macroblocks*. Luminance is represented with a higher resolution than chrominance, with a 4:2:0 ratio of sampling. Let us denote with $f_n(i, j)$, the pixel in the position denoted by the Cartesian coordinates (i, j) of the frame in the sequence n, as is shown in Figure 7.1.

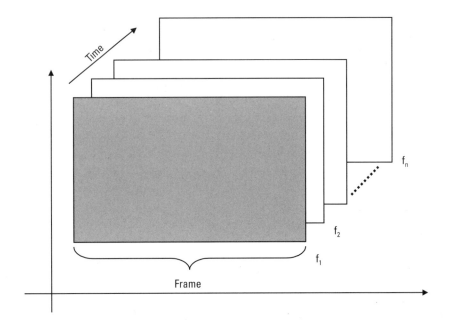

Figure 7.1 Video representation.

A general system for digital image compression consists of the following operations, which may be combined with each other:

- Quantization;
- Redundancy removal;
- Lossless encoding.

Before describing these steps in more detail, the issue of the quality criterion definition is addressed.

7.3.1 Performance Metrics

For continuous media services, like audio and video, as well as graphics, the relevant quality criterion is subjective. Subjective quality can be approximated by attributes such as frame rate and resolution (for video), bandwidth (for audio), and delay (for both video and audio). It is also measured by other factors, which are more difficult to characterize, such as the perceptual impact of artifacts introduced in the process of decompression, by information corrupted or discarded in the service (i.e., in the compression), and in packet losses and corruption. In most of the video coding process, losses are manifested as blocking, blurring, ringing, and mosquito noise. It is difficult to derive an objective measure of the perceived audio and video quality that encapsulates the effect of coding artifacts as perceived by the end user. However, it is fundamental to have objective performance metrics to optimize algorithms' parameters or to compare different algorithms without setting up time-consuming experiments that can be used only as a final test of the designed algorithm. There is a significant ongoing research in this area, attempting to derive perceptual metrics for video quality. One popular metric, which is currently used to rank the distortions introduced by video coding algorithms, is the average peak signal-to-noise-ratio (PSNR) given by [15]

$$PSNR = 10 \log_{10} \frac{\sum_{i=1}^{p} \sum_{j=1}^{q} (255)^2}{\sum_{i=1}^{p} \sum_{j=1}^{q} \left(f_n(i,j) - \tilde{f}_n(i,j) \right)^2} \qquad (7.1)$$

where $\tilde{f}_n(i,j)$ represents the compressed video information and 255 is the maximum value of a pixel in 8-bit representation. PSNR compares the maximum possible signal energy to the noise energy, and it has been shown to result in a higher correlation with the subjective quality perception of images than the conventional SNR given by

$$SNR = 10 \log_{10} \frac{\sum_{i=1}^{p} \sum_{j=1}^{q} f_n(i,j)^2}{\sum_{i=1}^{p} \sum_{j=1}^{q} \left(f_n(i,j) - \tilde{f}_n(i,j) \right)^2} \qquad (7.2)$$

Nonperceptual distortion criteria, such as PSNR, were found to be reasonably reliable for higher bit rates (high-quality applications), but they do not correlate properly with the perceived quality at lower rates and they fail to guarantee preservation of important perceptual qualities in the reconstructed images despite the potential for a good SNR. Ongoing research is attempting to produce models of the human visual system (HVS) and to develop a video quality metric by utilizing these models [16]. Based on psycho-visual experiments, a closed form expression of the contrast sensitivity function (CSF), which include properties of the human visual system, has been developed in [17]. A better prediction of perceived quality can be achieved by incorporating this CSF model in a distortion measure. Several coding schemes using models of human vision have been proposed [18–22]. Results that use a perceptual-based model in the optimization of a coder have also been reported in [23].

7.3.2 Quantization

If an image is represented by an analog waveform, the first step to digitally compress it is the digitization in space (sampling) and in amplitude (quantization). Quantization consists in mapping a continuous-valued amplitude into a finite set of approximating values. This operation is nonlinear and not invertible: it is lossy. The analog/digital conversion can operate on individual pixels (scalar quantization, SQ) or groups of pixels (vector quantization, VQ). Quantization can include throwing away some of the components of the signal decomposition involved in the redundancy removal step. The quality of the approximation may be improved by reducing the size of the steps, thereby increasing the number of allowable levels. Very small steps will not allow the human ear or the eye to distinguish the original from the quantized signal. To give the reader an idea of the number of quantization levels required in a practical system, 256 levels can be used to obtain the quality of commercial color TV, while 64 levels give only fairly good color TV performance.

7.3.3 Redundancy Removal

This step refers to all the techniques that are aimed at removing the temporal and spatial redundancy.

7.3.3.1 Motion Compensation

Motion compensation is the stage of video coding that reduces the temporal redundancy of the video signal. Exploiting the fact that two consecutive image frames, f_n and f_{n-1}, typically do not exhibit dramatic scene changes, motion compensation techniques approximate or predict the current frame by the previous one and encode the difference $e_n = f_n - f_{n-1}$. Most areas of this difference frame are flat, having values close to zero, and the variance is significantly lower than that of the original frame. Therefore, this reduced-variance difference signal, often referred to as motion compensated error residual (MCER), can be represented by a lower number of bits than f_n. The MCER can be encoded with the required distortion using a variety of techniques [15]. At the receiver the original frame $f_n = e_n + f_{n-1}$ has to be reconstructed by the approximation of the previous frame $\tilde{f}_{n-1}$ and the decoded quantized error signal $\bar{e}_n$, as follows:

$$f_n = \bar{e}_n + \tilde{f}_{n-1} \tag{7.3}$$

Prediction techniques suffer from the propagation errors, unless countermeasures are employed.

The described motion compensation technique, also referred to as *frame-differencing*, is not able to track complex motion trajectories where different objects move in different directions. More efficient, but also more complex, motion compensation techniques have been developed. In an implementation of block-based MC, which is used quite often, for each macroblock in the current frame (reference block), a matching macroblock is found in the past frame, which minimizes the pixel intensity differences between the two macroblocks. The two-dimensional vector that contains the location of the target macroblock (i.e., the found matching macroblock from which it is assumed that the reference macroblock has originated) is called motion vector (MV). The current block is predicted by the previous block, motion translated using the MV, and is subtracted from the current block before being encoded and transmitted, as is shown in Figure 7.2.

7.3.3.2 Spatial Redundancy Removal

A simple example of spatial redundancy removal techniques is the differential pulse code modulation (DPCM). Instead of coding the raw pixel values at each location, this technique encodes the difference between the raw pixel value and the prediction based on the neighboring pixel values. When there is a high degree of spatial similarity between pixels in a local neighborhood, as usually in images, this technique is very effective. Other techniques that are used to reduce the spatial redundancy are described in the rest of the section. Such techniques involve multiresolution decomposition.

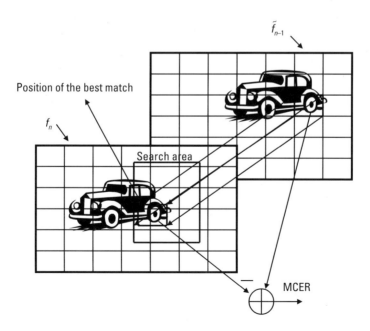

Figure 7.2 Motion compensation principle.

7.3.3.3 Pyramid Coding

Pyramid coding codes two versions of the signal [24]: first, lowpass filtering and subsampling derive a coarse version of the original signal; then, the original signal is predicted, based on this first coarse approximation, and the prediction error or difference signal, which represents the detail features, is evaluated. By iterating the scheme on the coarse image, a sequence of lower and lower resolution images of geometrically decreasing sizes (pyramid) is achieved. The process is diagrammatically shown in Figure 7.3.

The information that has to be represented efficiently is the lowest resolution version and the various difference images. The original image and the received image for different levels of the pyramid are shown in Figure 7.4, for comparison.

7.3.3.4 Transform Coding

Transform coding is the most pervasive technique for lossy compression of audio, images, and video [25]. The concept of transform coding can be summarized as follows [26]: instead of quantizing the single components of a discrete-time, continuous-valued vector source with correlated components, the source vector is linearly transformed and the *transform coefficients* are quantized. Transform coding is an inherently suboptimal source coding technique because it uses a *scalar quantizer*. An appropriately designed vector quantizer outperforms a scalar quantizer in terms of trade-off rate-distortion [27, 28], but it has much higher complexity. Therefore, transform coding is often called a low-complexity alternative to vector quantization.

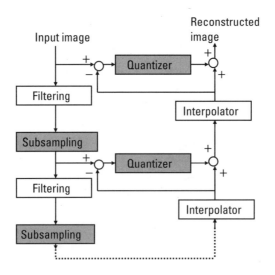

Figure 7.3 Principle scheme of pyramid coding.

Original image (a)

(b)

(c)

(d)

Figure 7.4 Comparison between the original image and the received image for different level of the pyramid, interpolated to original size for display: (a) original image; (b) lowest resolution; (c) resolution level #1; and (d) resolution level #2.

A popular transform coding is the discrete cosine transform (DCT), used by the Joint Photographic Experts Group (JPEG) image compression standard for lossy compression of images. Since the JPEG standard is so widely used, DCT is often referred to as the JPEG-DCT. Its popularity is mainly due to the fact that it achieves a good data compression—that is, it concentrates the information content in a relatively few transform coefficients. JPEG-DCT is a transform coding method comprising four steps. The source image is first partitioned into nonoverlapping subblocks of size 8×8 pixels in dimension. Then each block is transformed from spatial domain to frequency domain using 2D DCT basis function. The resulting frequency coefficients are quantized and finally transferred to a lossless entropy coder. DCT coefficient can be quantized lossily according to some human visual characteristics. Therefore, the JPEG image file format is very efficient in compressing image. This makes it very popular, especially in the World Wide Web. At high compression ratios, however, this scheme leads to a noticeable blocking artifacts-artificial discontinuities that often appear between the boundaries of the subblocks [29]. Various methods to cope with this problem without increasing the bit rate have been proposed, including lowpass filtering, constrained optimization [30] and projection on convex sets (POCS) [31]. Subband and wavelet coding are characterized by less quality degradation due to blocking artifacts.

7.3.3.5 Subband and Wavelet Coding

Subband coding is a specific form of transform coding, which does not suffer from blocking artifacts. The principle scheme of a two-channel subband coding is shown in Figure 7.5.

A filters bank calculates an orthogonal expansion of the input signal and the expansion coefficients are quantized [32, 33]. The analysis filter with impulse response $f_0(.)$ is the lowpass band filter, while the highpass band filter has impulse response $f_1(.)$. Subband coding has been used in speech compression [34] and images [35].

A significant amount of research has demonstrated the benefits of an alternative subband decomposition based on the *wavelet transform* [36–38]. The wavelet transform can be interpreted as a subband decomposition of the signal into a set of overlapping frequency channels having the same bandwidth on a logarithmic scale. An example of wavelet decomposition is shown in Figure 7.6.

The wavelet is able to both compact in frequency the energy into a small set of low-frequency coefficients and to compact spatially the energy into a small set of localized high-frequency coefficients, with the exact extent of the localization depending on the spatial support of the wavelet filters. One of the most efficient wavelet-based image coders for a noiseless channel is the zerotree algorithm developed by Shapiro [39] and further improved by Said and Pearlman [40].

7.3.3.6 Perceptual-Based Coding

Perceptual-based coding algorithms attempt to discriminate between signal components that are and are not detected by the human senses [17]. They attempt to remove redundant as well as the perceptually less significant information. Based

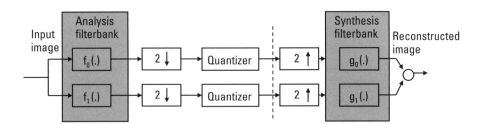

Figure 7.5 Principle scheme of two-channel subband coding.

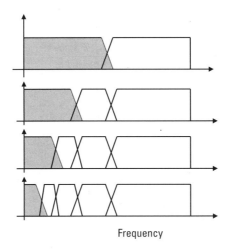

Frequency

Figure 7.6 Dyadic wavelet decomposition.

on psychophysical masking phenomena, detection thresholds are established according to a just-noticeable distortion (JND) and minimally noticeable distortion (MND) criterion. Perceptual coding aims at (1) hiding coding distortion beneath the detection thresholds and (2) augmenting the classical coding paradigm of redundancy removal with elimination of perceptually irrelevant signal information.

The first step in designing an image coder with a perceptual distortion control is to select a perceptual distortion metric. In the context of perceptual image compression, it is especially interesting to determine the bit rate at which a distortion becomes just perceptible. This perceptual lossless point constitutes a natural target for high quality, or for perceptually lossless image coding, which is important in applications where loss in quality cannot be tolerated, such as medical imaging. The perceptual lossless point can also be used as a benchmark for performance comparison with other coders that use perceptual distortion metrics. A JND-based model, by definition, provides a natural way of determining the perceptually lossless point at the pixel or the transform coefficient level. While the JND thresholds provide a localized measure of the noise threshold for a single subband coefficient, a perceptual distortion metric must also account for the spatial and spectral summation of individual quantization errors. In the coder proposed in [17], visual masking thresholds are derived by exploiting the human visual masking properties. The derived masking thresholds are used to adapt the quantizer reconstruction levels to meet the desired target perceptual distortion. Since the locally adaptive coder is optimized to adapt the quantization level to the local JND thresholds, and not to minimize MSE, there is no gain in terms of the MSE-based PSNR between the proposed coder and a nonadaptive coder. However, if the images are compared in terms of the perceptual distortion measure, the proposed locally adaptive coder outperforms a nonadaptive DC-tuner coder [41, 42] at all rates [17].

7.3.4   Lossless Encoding

To improve the coding efficiency, lossless compression, also called *entropy encoding* or *invertible encoding*, is used. Lossless encoding processes a bit stream without knowledge of the underlying signal semantics, which reduces the source data rate by exploiting the statistical distribution of the transmitted bits. The idea is to assign codewords with a few bits to likely symbols and codewords with more bits to unlikely symbols, so that the average number of bits is minimized. Variable-length codes, such as Huffman coding, Lempel-Ziv, and arithmetic codes are examples of lossless encoding.

7.4   Video Streaming

Audio and video streaming is a one-way transmission process where the data is viewed as it is received. Streaming requires that the data be transmitted at least at the rate that is used for playing them back on the client platform and that transmission latency does not increase significantly during playback. This transmission mode for stored video over the Internet has several advantages:

- Playback can be started quickly.
- Clients need only a small amount of storage space, which it is uses to buffer the transmission against relatively small changes in latency (jitter) caused by transmission delays and network congestion.
- A streaming multimedia server can tailor data it transmits to the available bandwidth and to the playback capabilities of the client workstation.
- If the user wants to stop watching the video, there has not been a waste of resources for an useless download.

Sending video over the Internet basically has two components: the compression of video data and the design of communication protocols. The former is presented in Section 7.3. In this section, further technologies and research needed to support streaming video over the Internet are highlighted [43]. First, the main features of well-established and emerging standards for streaming video are presented. Then, technologies proposed to fulfill some key requirements of this application are introduced.

7.4.1 Standards for Video Streaming

In this section, the main features of current video compression standards are recalled. The following subsections present other streaming-related standards such as IP multicast and the Real-Time Streaming Protocol (RTSP).

7.4.1.1 Video Compression Standards

There are two important international standard bodies that have developed or are developing low bit rate video compression standards: International Telecommunication Union (ITU) and the Moving Pictures Experts Group (MPEG) committee of the International Standardization Organization (ISO). ITU has developed two standards for videoconferencing and videophone applications: H.261 and H.263. H.261, which was approved in 1990, was intended for use on ISDN lines at some integer multiple of 64 Kbps. The new video codec H.253 was finalized in 1995 and allows operation over the Public Switched Telephone Network (PSTN), at bit rate lower than 64 Kbps. More information about H.261 can be found in [15] and T. Turletti's report on an H.261 video codec for use on the Internet [44, 45].

ISO/MPEG has developed the following well-known compression standards:

- MPEG-1 is intended for the storage and retrieval of video and audio on compact disc (i.e., at a bit rate of about 1.5 Mbps) [46–48].

- MPEG-2 is intended for high-quality broadcast applications, including entertainment, remote learning, electronic publishing, and high-quality video applications such as digital versatile disks (DVDs) [49].

- MPEG-3 was intended to address high-definition television, but this effort was folded into MPEG-2.

- MPEG-4 is intended to address wireless interactive multimedia coding and transmission [50].

- MPEG-7 addresses standardization of multimedia content descriptions so as to enable applications such as content-based remote and local browsing [51].

Work on the new standard MPEG-21 "Multimedia Framework" started in June 2000 [51].

Those standards generally assume a reliable transport mechanism and hence discard corrupted data and attempt to mask the discarded information. Other standards—those designed for a very unreliable transport (such as voice compression in digital cellular telephony [52] and video compression designed for multiple access wireless applications [53])—use corrupted data as if it were error-free and minimize the subjective impact of errors.

Most of the mentioned video coding standards, such as MPEG-1, MPEG-2, H.261, and H.263, are block-based coders where the image is first partitioned in nonoverlapping blocks. They all include the following four stages of processing: motion compensation, DCT coding, scalar quantization, and lossless coding. These stages have been previously described.

MPEG-1 and MPEG-2 are successful standards that have given rise to widely adopted commercial products, such as CD-interactive digital audio broadcasting and digital television. Limitations of these standards are mainly due to the block-based data representation model, in which a frame is divided in rectangular blocks. Block-based coders make very few assumptions about the input data and hence work very well over a variety of input video data and across a wide range of bit rates. The interactivity, however, which is one of the key requirement in the newer applications such as Internet video, DVD, and video games, cannot be provided. In fact, in order to provide inter-activity it should be possible to decompose the scene into objects, which can be, for instance, deleted by the given scene or introduced on a given background.

Other approaches have reached maturity such as object-based coding and model-based coding [1]. The objects-based representation approach, where a scene is modeled as a composition of objects, both natural and synthetic, with which the user may interact, is at heart of the MPEG-4 technology [50, 54]. An object can be: a walking person, a moving car, or the ball on the foot of a soccer player. Since appropriate processing can be applied for each object, these techniques achieve efficient compression, but principally they can support content-based functionalities, such as the ability to selectively code, decode, and manipulate specific objects in the video streams. For example, users may specify objects to be viewed without the background, or a background scene without objects in the foreground. The position data of each object can be coded inde-pendently from the video data. Because of this, video processing for program production and editing can be done simply by changing the position data. In MPEG-4, the coding principle of the block-based coders is extended by the so-called video object plane formation. The video object is circumscribed by a bounding rectangle, which is determined so that the number of image mac-roblocks to be coded is as small as possible. A macroblock that lies completely outside the image object is labeled a *transparent macroblock* and can be skipped during texture coding. A macroblock that is situated completely inside the image object is called object macroblock and can be coded straightforwardly using motion compensation and block-based DCT. The third class of mac-roblocks includes those macroblocks that only partially belong to the video object, and these are called *boundary macroblocks*. Those macroblocks require completely different coding techniques. There are two basic approaches for coding these blocks:

1. *Padding methods:* The partially filled image block is extended to its full size by padding the missing data. The padded block is then coded using standard DCT.
2. *Shape adaptive transform:* The boundary blocks are coded using a shape-adaptive transform [54].

Which method should be used for boundary block coding depends on the application and data rate envisaged. For mobile communications and object-based video, retrieval of a low data rate seems to be preferable to use padding techniques since they have low computational complexity and can be easily implemented in hardware. A shape-adaptive transform seems to be a suitable choice for applications that require a high video quality and allow for some more complex hardware.

MPEG-4 version 1 was approved by MPEG in December 1998; version 2 was frozen in December 1999. After these two major versions, more tools were added in subsequent amendments that could be qualified as versions. It is ITU H.263 compatible in the sense that a H.263 bit stream is correctly decoded by MPEG-4.

If compared with other existing compression standards, MPEG-4 also provides:

- Ability to operate at much higher rates (tested up to 6 MB);
- Improved coding efficiency;
- More flexible scalability mechanisms (see Section 7.4.2);
- Improved error resilience for transmission over noisy communication channels.

An in-depth presentation of MPEG-4 can be found in [50–55]. In response to growing need on a video-coding standard for streaming video over the Internet, MPEG-4 has been actively defining profiles for this application such as fine granularity scalability (FGS), which is described in Section 7.4.2.

7.4.1.2 IP Multicast

A majority of streaming video today uses unicast, in which a point-to-point connection is set up between a server and client. There is an increasing push to move towards multicast, which allows the deployment of multimedia applications on the network while minimizing their demand for bandwidth. In fact, when unicast is used, a separate copy of the data is sent from the source to each client that requests it. When the same data needs to be sent to only a portion of the clients on the network, the unicast approach wastes bandwidth by sending multiple copies of the data. In case of multicast, a single copy of the data is sent to those clients who request it and multiple copies of data are not sent across the network.

The original Internet design, while well suited for point-to-point applications like e-mail, file transfer, and Web browsing, fails to effectively support large-scale content delivery, like streaming multicast. Several features are being implemented for TCP/IP multicasting, including upgrades to IP itself (see Chapter 3). For example IPv6 includes different addressing for unicast, anycast (communications with the closest member of a device group), and multicast transmission. Standards for IP multicasting over IPv4 are provided by several protocols including the Internet Group Management Protocol (IGMP), Protocol Independent Multicast (PIM), and Distance Vector Multicast Routing Protocol (DVMRP). Mbone has been an initial experimental application that uses multicasting. Mbone is a best-effort network; therefore, in order to provide QoS guarantees, it requires application-layer QoS mechanisms for video communications, such as rate and error control mechanisms (see Section 7.4.3). However, the implementation of such rate and error control functions is considerably more challenging for multicast than unicast. This is because in a unicast setting, the needs of only a single receiver have to be addressed, whereas in a multicast setting, it becomes necessary to cater simultaneously to the potentially conflicting needs of multiple recipients. For example, in unicast systems the source usually adapts the transmission rate according to a feedback from the receiver. In a multicast transmission, by requiring feedback information from all receivers, the source could be overwhelmed by feedback messages from the receivers. Furthermore, by sending the same single stream to all receivers, this adaptation scheme cannot accommodate the heterogeneity of channel conditions and user preferences. The layered multicast framework offers a flexible solution for scalable video [56, 57] that produces embedded bit streams, as explained in Section 7.2.2. In [7] the layered FEC is proposed for layered multicast systems. By organizing FEC into multiple layers, receivers can obtain different levels of protection commensurate with their respective channel conditions.

7.4.1.3   Transport Protocols

TCP/IP and UDP suite provide the basis of Internet. The main difference between UDP and TCP is that TCP provides reliable transmission and uses retransmission to recover lost packets. Since these retransmissions introduce delays that are not acceptable for streaming applications with stringent delay requirements, UDP is typically employed as the transport protocol for video stream. Since UDP does not guarantee packet delivery, the receiver needs to rely on upper layer protocols to detect packet loss. Real-Time Transport Protocol (RTP) and Real-Time Control Protocol (RTCP) are examples of layer 4 protocols that run on top of UDP/TCP and are designed for end-to-end, real-time delivery of data such as video and voice.

The functionalities provided by RTP include content identification, time reconstruction, loss detection, and security. It can deliver data to one or more destination with a limit on the delay variations. RTP does not guarantee QoS or reliable delivery. RTCP is the control protocol designed to work in conjunction with RTP. The primary function of RTCP is to provide feedback to an application about the quality of data distribution. The feedback is in the form of a report that can be sent by the source or the receiver and contains information such as: (1) fraction of the lost RTP packets since the last report, (2) cumulative number of lost packets since the beginning of reception, (3) packet interarrival, and (4) delay since receiving the last sender's report. Based on this information, the sender can adjust its transmission rate (see Section 7.4.3); the receiver can determine whether congestion is local, regional, or global; and network managers can evaluate network performance for multicast distribution.

Finally, the reader can refer to Chapter 4 for an overview of the strategies exploited to improve TCP/IP protocol over wireless links in the presence of services with tight transport/delay constraints.

7.4.2 Scalability

Scalability of the coding rate used is fundamental to gracefully cope with the bandwidth fluctuations of the Internet [1]. As it has been already outlined, Multiresolution decomposition is a key approach to provide scalability. Decoding multiple substreams produces pictures with degraded quality, or smaller size, or lower frame rate. The scalabilities of quality, image size, and frames rates are called SNR, spatial and temporal scalability, respectively, and are the basic scalable mechanisms. Most of the existing video compression standards include these basic mechanisms. There can be combinations of these basic mechanisms, such as spatio-temporal scalability [58]. In such a layered scalable coding technique, a video sequence is coded into a base layer and an enhancement layer. The enhancement layer is either entirely transmitted/received/decoded or it does not provide any enhancement at all to the video quality. Through these techniques, the video quality is optimized at a given bit rate. The objective of video coding over the Internet is to optimize the video quality over a given range of bit rates. The bit stream should be partially decodable at any bit rate within the bit rate range to reconstruct a video signal with the optimized quality at that bit rate. To provide more flexibility in meeting different demands of streaming, a new scalable coding mechanism, called fine granularity scalability (FGS) was proposed to MPEG-4 [59–61]. The major difference between the layered coding and FSG coding is that, although the coding technique encodes a video sequence into two layers, the enhancement bit stream can be truncated into any number of bits within each frame, in order to provide partial enhancement proportional to the number of bit decoded for each frame. This is achieved by encoding the base layer using nonscalable coding, and the enhancement bit stream using bit-plane coding of the DCT coefficients. Bit-plane is a coding technique that uses embedded representations (see Section 7.2.2) [62]. A variation of FGS is progressive fine granularity scalability (PFGS). Unlike FGS that has only two layers, PFGS could have more than two layers. The essential difference is that FGS only uses the base layer as a reference for motion prediction, while PFGS uses multiple layers as reference to reduce the prediction error, resulting in higher coding efficiency.

7.4.3   Applications-Layer QoS Mechanisms

The current best-effort Internet does not offer any QoS guarantees to streaming video over the Internet. To avoid congestion and maximize video quality in presence of packet loss, application-layer QoS control techniques are employed by the end systems, which do not require any QoS support from the network. These include congestion control and error control. Congestion control usually takes the form of rate control [63]: rate control attempts to minimize the possibility of network congestion by matching the rate of video stream to the available network bandwidth.

Error control mechanisms attempt to minimize the visual impact of lost packets at the destinations. FEC and retransmissions are well-known error control mechanisms (see Chapters 4 and 5). Retransmission is usually not applied to recover the lost packets in real-time video because the introduced delay could be excessive. In some cases, however, a retransmission approach (called delay-constrained retransmission) may be possible. For Internet applications, channel coding is used in terms of block codes. The video stream is chopped in segments of k packets. For each segment a block code (e.g., Tornado code [1]) is applied to the k packets to generate an n-packet block. Error control mechanisms are also applied.

FEC can be used in a layered multicast so that each client can individually trade off latency based on its requirements. Examples of such an FEC-protected multicast include hierarchical FEC [57] and a receiver-driven layered multicast [56].

7.4.3.1   Error Resilient Encoding

Error resilient encoding is used to enhance the robustness of compressed video to packet loss. As already shown, most of the low bit rate coding standards use predictive coding techniques, such as block-motion compensation and entropy encoding to achieve high compression. Variable-length codes, such as Huffman codes, render the compressed stream very sensitive to channel errors. As a result, the video decoder, which is decoding the corrupted video stream, loses synchronization with the encoder. Moreover, predictive techniques quickly propagate the effects of channel errors across the video sequence rapidly degrading video quality, unless proper countermeasures are taken. The standardized error resilient encoding schemes include data recovery, resynchronization marking, and data partitioning. Resynchronization marking is a countermeasure for the propagation error of predictive techniques. Resynchronization markers are introduced in the bit stream at various locations. When the decoder detects an error, it can then hunt for this resynchronization marker and regain synchronization. After detecting the error in the bit stream and resynchronizing to the next resynchronization marker, the decoder now has to isolate the data in error that is in the macroblocks between the two resynchronization markers. Usually between two resynchronization markers, the motion and DCT data for each macroblock are coded all together. Hence, when the decoder detects an error, whether the error occurred in the motion or DCT part, all the data in the video needs to be discarded. In MPEG-4, the *data partitioning* mode partitions the data within the video packet into a motion part and a texture part separated by a unique motion boundary marker (MBM). The advantage of this partitioning is twofold: (1) a more stringent check on the validity of motion data can be provided, since the MBM has to be taken at the end of the decoding of motion data, in order to be valid; and (2) in case there is an undetected error in the motion and texture, but we do not end in the correct position for the next resynchronization marker, we do not need to discard all the motion data since we can save the motion data as validated by the detection of the MBM.

These mechanisms are targeted at error-prone environments like wireless channels. A promising error-resilient technique for robust Internet video transmission is the multiple description coding (MDC) [7]. MDC is suitable over communications systems with multiple channels (i.e., a source connected to the destination by two parallel channels). According to MDC, a raw video sequence is compressed into multiple streams (referred to as descriptions) as follows: each description provides acceptable visual quality; more combined description provide better visual quality. In this way, even if a receiver gets only one description (other descriptions being lost), it can still reconstruct video with acceptable quality. If a receiver gets multiple descriptions, it can combine them together to produce a better reconstruction. However, this technique will reduce the compression efficiency if compared with a conventional single description coder. Further investigation is needed to find a satisfactory trade-off between compression efficiency and reconstruction quality.

7.4.3.2 Error Concealment Encoding

Error concealment is performed by the receiver with the aim at minimizing the degradation of the picture quality by utilizing the correctly decoded data in masking the effects of residual errors (those not removed by the error protection scheme). There are two approaches for error concealment: spatial and temporal interpolation. Spatial interpolation reconstructs a missing piece in a frame from its adjacent (presumably nonmissing) regions in the same frame. It has poor performance in block-based coders, because several consecutive lines are damaged when the packet is lost. In temporal interpolation, lost data is reconstructed by data from the previous frame. Several error concealment schemes have been proposed in recent years [64].

7.4.4   Joint Source/Channel Coding

An important information theoretic result from Shannon [65], known as the *separation principle*, states that it is possible to separate the design of a source compression/decompression scheme and a channel coding/decoding scheme without loss of performance, as long as the minimum achievable source coding rate of the source is strictly below the capacity of the channel and under the following assumptions:

- Use of very long block length for both source and channel codes;
- Availability of arbitrarily large computational resources;
- No delay constraints.

In several practical situations such conditions are not met [66] and the separation principle does not hold anymore, particularly for real-time services like audio and video:

- The focus is on minimizing the delay, and thus, the block sizes need to be bounded for source coding;
- Complexity is a major factor in compression and channel coding systems;
- The only meaningful criterion of quality is subjective, and such a quality criterion falls outside the scope of the separation theorem.

Therefore, in these cases substantial gains can be achieved in performance of reduction complexity for a given subjective quality through a joint source/channel code design [67–70]. The UEP is an example of joint source-channel approach. An area where the joint source-channel coding principle had an impact is communicating over heterogeneous networks [2, 3], particularly in the case of multicast in heterogeneous environment. A multicast transmission can be conceptualized as existing on a Multiresolution tree. Each user can reach as many levels of the Multiresolution tree as is possible, given its access capabilities.

7.5 VoIP

Internet telephony uses Voice over IP (VoIP) technology over the public Internet or a private IP network to carry voice calls, either partially or completely bypassing the PSTN [71]. Initially, VoIP was used by computer hobbyists to place long-distance calls over the Internet for the price of a local call. Sound quality was—and often still is—quite poor. The picture changed when companies started to see the commercial advantages of this application. Enterprises and carriers can provide various services using VoIP, including enterprise toll bypass, IP-based IntereXchange Carrier (IXC), long-distance service, and IP-based local telephony.

VoIP technology requires the following:

1. To convert voice and fax calls from analog to digital form. This is also done in the PSTN for interoffice transmission, but in a less efficient way. VoIP codecs allow voice transmission at 8 Kbps, while a phone line requires 64 Kbps;

2. To encapsulate this digitally encoded voice in IP packets using RTP (See Section 7.4.1), which runs over the UDP.

3. The packet are carried over an IP network, which ideally incorporate QoS features to give higher priority to delay-sensitive traffic such as RSVP or other mechanisms;

4. A VoIP system also must incorporate call setup and control features.

These and other functions often are based on ITU H.323 Internet telephony standard, which is the most widely known set of standards governing VoIP. It comprises subrecommendations for specific aspects of the interface between the PSTNs and IP networks, such as signaling and voice encoding. The IETF has also addressed VoIP with its Session Initiation Protocol (SIP) set. SIP handles the creation, modification, and termination of calls.

The prevalent speech coders for VoIP are hybrid coders, which combine the attractive features of waveform coders and vocoders. Waveforms coders aim at reproducing the analog waveform as accurately as possible, including the background noise. The vocoders build a set of parameters to describe the waveform to be reproduced. These parameters are sent to the receiver that uses them to drive a speech production model. The quality of vocoders is not good enough for use in telephony systems, while the waveform coders operate at high bit rate (64 Kbps). Hybrid vocoders can work at low bit rate (4–16 Kbps) and use analysis-by-synthesis (AbS) techniques [71].

Like in the case of video over the Internet, the problems to be faced are both in the compression techniques and communication protocols, and most of the solutions described for video over Internet apply also to VoIP.

7.6 Conclusions

One of the main challenges related to the delivery of high-quality video and audio in future communication systems, or what makes all the other technical issues more challenging, is the heterogeneity of the communication network. For example, the Internet includes clients and servers with widely different capacities as well as diverse and dynamic network connections between them. This diversity causes the amount of resources available between video servers and clients to vary, both from network to network, and dynamically at a single network. If insufficient resources are available anywhere along the video pipeline, quality rapidly degrades to unacceptable levels. One approach for avoiding this outcome is to have the applications use reservations to ensure enough resources are available (RSVP) (see Chapter 3). Furthermore, due to the large variety of existing network technologies, it is most likely that hybrid networks are used to support multimedia services. Different networks have different characteristics. Optimizing the performance over such heterogeneous networks requires more flexible and scalable video coding that still has high compression efficiency and low complexity. One promising solution for meeting all these conflicting requirements is to combine/integrate several video coding techniques in a layered structure [72]. The video information is decomposed in layers in different dimensions: bit rate dimension, delay tolerance dimension, and error resilience dimension. The core video information is the most visually significant data requiring the minimum bandwidth, the least error resilient, and the lowest delay. The core layer can only carry key frames with an aggressive compression and it has no adaptability to any fluctuation. Adding layers/levels along one or more dimensions increases adaptability. Moreover, a new design approach is needed that jointly considers data compression techniques and communication protocols [3].

References

[1]   Bull, D., N. Canagarajan, and A. Nix, *Mobile Multimedia Communications*, Academic Press, 1999.

[2] Chen, J., K. J. Ray Liu, "Joint Source-Channel Multistream Coding and Optical Network Adapter Design for Video over IP," *IEEE Trans. on Multimedia*, Vol. 4, No. 1, March 2002.

[3] Servetto, S. D., and K. Nahrstedt, "Broadcast Quality Video over IP," *IEEE Trans. on Multimedia*, Vol. 3, No. 1, pp. 162–173, March 2001.

[4] Civanlar, M. R., A. Luthra, S. Wenger, "Introduction to the Special Issue on Streaming Video," *IEEE Trans. on Circuits and Systems for Video Technolgy*, Vol. 11, No. 3, March 2001.

[5] Talluri, R., "Error Resilient Video Coding in the ISO-MPEG-4 Standard," *IEEE Commun. Mag.*, June 1998, pp. 112–119.

[6] Talluri, R., et al., "A Robust, Scalable, Object-Based Video Compression Technique for Very Low Bit Rate Coding," *IEEE Trans. on Circuits and Systems for Video Technolgy*, Vol. 7, No. 1, Feb. 1997, pp. 221–233.

[7] Wang, Y., and S. Lin, "Error Resilient Video Coding Using Multiple Description Motion Compensation," *IEEE Trans. on Circuits and Systems for Video Technology*, Vol. 13, No. 6, June 2002, pp. 438–452.

[8] Poor, H. V., G. W. Wornell, and V. Poor, *Wireless Communications: Signal Processing Perspectives*, Englewood Cliffs, NJ: Prentice-Hall Signal Processing Series, April 1998.

[9] Mallat, S., "A Theory of Multiresolution Signal Decomposition: The Wavelet Representation," *IEEE Trans. Patt. Anal. Mach. Intell.*, Vol. 11, No. 7, July 1989, pp. 674–693.

[10] Hagenauer, J., "Rate-Compatible Punctured Convolutional Codes (RCPC codes) and Their Applications," *IEEE Trans. Commun.*, Vol. 36, April 1988, pp. 389–400.

[11] Kim, Y. H., and J. Modestino, "Adaptive Entropy Coded Subband Coding of Images," *IEEE Trans. Signal Processing*, Vol. 39, No. 1, Jan. 1992, pp. 31–48.

[12] Modestino, J., D. G. Daut, and A. Vickers, "Combined Source-Channel Coding of Images Using Block Cosine Transform," *IEEE Trans. Commun.*, Vol. COM-29, Sept. 1981, pp. 1261–1274.

[13] Karlsson, G., and M. Vetterli, "Packet Video and Its Integration into Network Architecture," *IEEE JSAC*, Vol.7, No. 5, June 1988, pp. 739–751.

[14] Nomura, M., T. Fujii, and N. Ohta, "Layered Packet-Loss Protection for Variable Rate Video Coding Using DCT," in *Proc. Sec. Int. Workshop on Packet Video*, Torino, Italy, Sept. 1988.

[15] Hanzo, L., P. Cherriman, and J. Streit, *Wireless Video Communications—Second to Third Generation Systems and Beyond*, Piscataway, NJ: IEEE Press, 2001.

[16] Watson, A. B., *Digital Images and Human Vision*, Cambridge, MA: MIT Press, 1993.

[17] Hontsch, I., and L. J., Karam, "Adaptive Image Coding with Perceptual Distortion Control," *IEEE Transactions on Image Processing*, Vol. 11, No. 3, March 2002, pp. 213–222.

[18] Nill, N. B., "A Visual Model Weighted Cosine Transform for Image Compression and Quality Assessment," *IEEE Trans. Commun.*, Vol. COM-33, July 1985, pp. 551–557.

[19]   Eggerton, J. D., and M. D. Srinath, "A Visually Weighted Quantization Scheme for Image Bandwidth Compression at Low Data Rate," *IEEE Trans. Commun.*, Vol. COM-34, Aug. 1986, pp. 840–847.

[20]   Saghri, J. A., P. S. Cheatham, and A. Habibi, "Image Quality Measure Based on a Human Visual System Model," *Opt. Eng.*, Vol. 28, No. 7, 1989, pp. 813–818.

[21]   Macq, B., "Weighted Optimum Bit Allocations to Orthogonal Transforms for Picture Coding," *IEEE J. Select. Areas Commun.*, Vol. 10, June 1992, pp. 875–883.

[22]   Daly, S., "Digital Image Compression and Transmission System with Visually Weighted Transform Coefficients," U.S. Patent 4 780 761, 1995.

[23]   Watson, A. B., "DCT Quantization Matrices Visually Optimised for Individual Images," in *Proc. SPIE: Human Vision, Visual Processing, and Digital Display IV*, Vol. 1913, pp. 202–216, San Jose, CA, Feb. 1993.

[24]   Burt, P. J., and E. H. Adelson, "The Laplacian Pyramid as a Compact Image Code," *IEEE Trans. Commun.*, Vol. 31, April 1993, pp. 532–540.

[25]   Docef, A., et al., "The Quantized DCT and Its Application to DCT-Based Video Coding," *IEEE Transactions on Image Processing*, Vol. 11, No. 3, March 2002, pp. 177–187.

[26]   Huang, J. J. Y., and P. M. Schultheiss, "Block Quantization of Correlated Gaussian Random Variables," *IEEE Trans. Commun. Syst.*, Vol. COM-11, Sept. 1963, pp. 289–296.

[27]   Lookabaugh, T. D., and R. M. Gray, "High Resolution Quantization Theory and the Vector Quantization Advantage," *IEEE Trans. Inform. Theory*, Vol. 35, Sept. 1989, pp. 1020–1033.

[28]   Na, S., and D. L. Neuhoff, "Bennett's Integral for Vector Quantizers," *IEEE Trans. Inform. Theory*, Vol. 41, July 1995, pp. 886–900.

[29]   Yu, A. C.-W., O. C. Au, and B. Zeng, "Removing of Blocking Artefacts Using Error-Compensation Interpolation and Fast Adaptive Spatial-Varying Filtering," *IEEE International Symposium on Circuits and Systems*, 2002, Vol. 5, pp. 241–244.

[30]   Chou, J., M. Crouse, and K. Ramchandran, "A Simple Algorithm for Removing Blocking Artefacts in Block-Transform Coded Images," *IEEE Signal Processing Letters*, Vol. 5, No. 2, Feb. 1998, pp. 33–35.

[31]   Yang, Y., N. P. Galatsanos, and A. K. Katsaggelos, "Projection-Based Spatially Adaptive Reconstruction of Block-Transform Compressed Images," *IEEE Trans. on Image Processing*, Vol. 4, No. 7, July 1995, pp. 896–906.

[32]   Vetterli, M., and J. Kovacevic, *Wavelet and Subband Coding*, Englewood Cliffs, NJ: Prentice-Hall, 1989.

[33]   Cosman, P. C., R. M. Gray, and M. Vetterli, "Vector Quantization of Image Subbands: A Survey," *IEEE Trans. Image Processing*, Vol. 5, No. 2, Feb. 1996, pp. 202–225.

[34]   Crochiere, R. E., S. A. Webber, and J. L. Flanagan, "Digital Coding of Speech in Subbands," *Bell Syst. Tech. J.*, Vol. 55, Oct. 1976, pp. 1069–1085.

[35]   Vetterli, M., "Multidimensional Subband Coding: Some Theory and Algorithms," *Signal Processing*, Vol. 6, Feb. 1984, pp. 97–112.

[36] Shapiro, J. M., "An Embedded Wavelet Hierarchical Image Coder," in *Proc. Int. Conf. Acoust., Speech, Signal Processing*, San Francisco, March 1992, pp. 657–660.

[37] Lewis, A. S., and G. Knowles, "Image Compression Using the 2-D Wavelet Transform," *IEEE Trans. Image Processing*, Vol. 1, April 1992, pp. 244–250.

[38] Antonini, M., et al., "Image Coding Using Wavelet Transform," *IEEE Trans. Image Processing*, Vol. 1, April 1992, pp. 205–220.

[39] Shapiro, J. M., "Embedded Image Coding Using Zerotree of Wavelet Coefficients," *IEEE Trans. Signal Processing*, Special issue on wavelet and signal processing, Vol. 41, Dec. 1993, pp. 3445–3462.

[40] Said, A., and W. A. Pearlman, "A New, Fast and Efficient Image Codec Based on Set Partitioning in Hierarchical Trees," *IEEE Trans. on Circuits and Systems for Video Technology*, Vol. 6, 1996, pp. 243–249.

[41] Watson, A. B., "DCTune: A Technique for Visual Optimization of DCT Quantization Matrices for Individual Images," in *Soc. Information Display Dig. Tech. Papers XXIV*, 1993, pp. 946–949.

[42] Watson, A. B., "DCT Quantization Matrices Visually Optimized for Individual Images," in *Proc. Human Vision, Visual Processing, Digital Display IV*, B. E. Rogowitz, (ed.), 1993, pp. 202–216.

[43] Wu, D., et al., "Streaming Video over the Internet: Approaches and Directions," *IEEE Trans. Circuits and Systems for Video Technology*, Vol. 11, No. 3, March 2001, pp. 282–300.

[44] Turletti, T., "A H.261 Software Codec for Videoconferencing over the Internet," Tech. Rep. 1834, INRIA, 06902 Sophia-Antipolis, France, Jan. 1993.

[45] Kenyon, N., and C. Nightingale, *Audiovisual Telecommunications*, London: Chapman and Hall, 1992.

[46] Le Gall, D. J., "The MPEG Video Compression Algorithm," *Signal Processing: Image Communication*, Vol. 4, No. 2, April 1992, pp. 129–140.

[47] Le Gall, D. J., "MPEG: A Video Compression Standard for Multimedia Applications," *Commun. ACM*, Vol. 34, April 1991, pp. 46–58.

[48] ISO/IEC Standard 11172, "Coding Moving Pictures and Associated Audio at Up to About 1.5 Mbits/s" (MPEG-1).

[49] ISO/IEC Standard 13818, "Generic Coding of Moving Pictures and Associated Audio" (MPEG-2).

[50] MPEG Video Group, "Overview of the MPEG-4 Standard," ISO/IEC JTCI/SC29/WG11 N2323, Dublin, Ireland, July 1998, http://mpeg.telecomitalialab.com/standards/mpeg-4/mpeg-4.htm.

[51] Koenen, R., "From MPEG-1 to MPEG-21: Creating an Interoperable Multimedia Infrastructure," ISO/IEC JTC1/SC29/WG11, Dec. 2001, http://mpeg.telecomitalialab.com/documents/from_mpeg-1_to_mpeg-21.htm.

[52] Natvig, J. E., S. Hansen, and J. De Brito, "Speech Processing in the Pan-European Digital Mobile Radio System," in *Proc. Globecom*, Dallas, TX, Vol. 2, Nov. 1989, pp. 1060–1064.

[53] Meng, T. H., et al., "Portable Video-on-Demand in Wireless Communication," *Proc. IEEE,* April 1995, pp. 659–680.

[54] Kaup, A., "Object-Based Texture Coding of Moving Video in MPEG-4," *IEEE Trans. Circuits and Systems for Video Technology,* Vol. 9, No. 1, Feb. 1999, pp. 5–15.

[55] http://leonardo.telecomitalialab.com/icjfiles/mpeg-4_si/7natural_video_paper/7-natural_video_paper.htm

[56] McCanne, S., V. Jacobson, and M. Vetterli, "Receiver-Driven Layered Multicast," in *Proc. ACM SIGCOMM'96,* Aug. 1996, pp. 117–130.

[57] Wai-tian Tan, and A. Zakhor "Video Multicast Using Layered FEC and Scalable Compression," *IEEE Trans. Circuits and Systems for Video Technology,* Vol. 11, No. 3, March 2001, pp. 373–386.

[58] Girod, B., U. Horn, and B. Belzer, "Scalable Video Coding with Multiscale Motion Compensation and Unequal Error Protection," in *Proc. Symp. Multimedia Communications and Video Coding,* New York, Oct. 1995, pp. 475–482.

[59] Li, S., F. Wu, and Y.-Q. Zhang, "Study of a New Approach to Improve FGS Video Coding Efficiency," ISO/IEC JTC1/SC29/WG11, MPEG99/M5583, Dec. 1999.

[60] Li, W., "Bit-Plane Coding of DCT Coefficients for Fine Granularity Scalability," ISO/IEC JTC1/SC29/WG11, MPEG98/M3989, Oct. 1998.

[61] Li, W., "Streaming Video Profile in MPEG-4," *IEEE Trans. Circuits Syst. Video Technol.,* Vol. 11, March 2001.

[62] Li, W., "Overview of Fine Granularity Scalability in MPEG-4 Video Standard," *IEEE Trans. on Circuits and Systems for Video Technology,* Vol. 11, No. 3, March 2001, pp. 301–317.

[63] Wu, D., Y. T. Hou, and Y.-Q. Zhang, "Transporting Real-Time Video over the Internet: Challenges and Approaches," *Proc. IEEE,* Vol. 88, Dec. 2000, pp. 1855–1875.

[64] Wang, Y., and Q.-F. Zhu, "Error Control and Concealment for Video Communication: A Review," *Proc. IEEE,* Vol. 86, May 1998, pp. 974–997.

[65] Shannon, C. E. "A Mathematical Theory of Communication," *Bell Syst. Tech. J.,* Vol. 27, 1948, pp. 379–423.

[66] Vembu, S., S. Verdu, and Y. Steinberg, "The Source-Channel Separation Theorem Revised," *IEEE Trans. Inform. Theory,* Vol. IT-41, Jan. 1995, pp. 44–54.

[67] Modestino, J., D. G. Daut, and A. Vickers, "Combined Source-Channel Coding If Images Using the Block Cosine Transform," *IEEE Trans. Comun.,* Vol. COM-29, Sept. 1981, pp. 1261–1274.

[68] Liu, F. H., and V. Cuperman, "Joint Source and Channel Coding Using a Non Linear Receiver," in *Proc. ICC,* Vol. 3, June 1993, pp. 1502–1507.

[69] Garret, M. W., and M. Vetterli, "Joint Source/Channel Coding of Statistically Multiplexed Real-Time Services on Packet Networks," *IEEE/ACM Trans. on Networking,* Vol. 1, No. 1, Feb. 1993, pp. 71–80.

[70] Goldsmith, A., "Joint Source/Channel Coding for Wireless Channels," in *Proc. IEEE VTC95*, 1995, pp. 614–618.

[71] Black, U., *Voice over IP*, Englewood Cliffs, NJ: Prentice-Hall, 1999.

[72] Chang, Y.-C., and D. G. Messerschmitt, "Adaptive Layered Video Coding for Multi Time Scale Bandwidth Fluctuations," available at http://di-vine.EECS.Berkeley.EDU/~messer/.

8

Personal Area Networks

8.1 Introduction

Zimmerman [1] introduced the term *personal area network* to denote a communication between proximal electronic devices by using the body as a conduit for information. This "body network" has been conceived as a support to the intelligent environment that the person "carries." The motivation for such a network comes from the fact that there is need for data exchange not only on large distances (which is commonly referred to as communications), but also between devices carried by one person or surrounding the person at a conversational distance. This concept of *personal* communication has been further developed and implemented through different technologies. The personal area network (PAN) is a new member of the telecommunication family. It will cover the personal space surrounding the person within the distance that can be covered by the voice. It will have a capacity in the range of 10 bps to 10 Mbps. Particular interest is received by the wireless PAN (WPAN), where the communication occurs through the wireless medium instead of through, for instance, the electric field of the human body as a conductor. Future WPANs should offer seamless, autoconfigured connections to available resources. WPANs represent an example of ad hoc networks, which provide the capability of establishing mutual on-demand wireless links with either little or no preconfiguration. So far, the research in ad hoc networking has not been bound to any particular technology. Although the concepts of ad hoc networking are much broader than the WPAN scenario, in the short term personal communications will mainly use products that apply these concepts. However, the design of ad hoc networks that takes into account the specific attributes of WPANs is a new area for future research. The main technical challenges are presented in this chapter. The chapter is organized as follows: the concept of PAN is introduced in Section 8.2 where possible applications and devices are presented; existing and emerging technologies for supporting these short-range communications are introduced in Section 8.3; technical challenges in designing WPANs are highlighted in Section 8.4; and conclusions are drawn in Section 8.5.

8.2 PAN Concept

PAN is a network solution that enhances our personal environment, either work or private, by networking a variety of personal and wearable devices (PDAs, Webpads organizer, hand computers, cameras, head mounted displays, etc.) that surround a person within the distance that may be covered by the voice, and providing communication capabilities within that personal space and with the outside world [2–4]. For the networking of proximate devices, PAN may use the wireless medium, the electric field of the human body as a conductor, the magnetic field, and so on. In particular, when it uses the wireless medium, it is referred to as wireless PAN. In the rest of the chapter we will mainly refer to WPAN technology. The WPAN forms a wireless "bubble" around the person, referred to as personal operating space (POS). A fundamental concept behind WPAN systems asserts that any time two WPAN-equipped devices get within approximately 10m of one another (i.e., their POSs intersect), they can form a spontaneous, just-in-time, disposable connection. A WPAN device may connect to wall repeaters to access to the Internet or it can be dynamically stretched to include access to sensors and actuators.

WPAN technology is a short-range wireless technology as is WLAN technology. However, WPAN and WLAN have complementary positions [5]: WPAN emphasizes low cost and low power consumption, usually at the expense of the transmission speed range and the maximum data rate. WLANs emphasize higher peak data rate and longer range at the expense of cost and power consumption. This concept is shown in Figure 8.1.

8.2.1 PAN Architecture

The general PAN network model is presented if Figure 8.2. The PAN is a network for the person, between persons, and between the person and the outside world. Therefore, the network architecture is a *layered architecture*, seamless to the user, where different layers cover the specific types of connectivity. The connectivity is enabled by incorporating different networking functionalities into different devices. Therefore, for the stand-alone PAN, the person should be able to address the devices within the POS independently of the surrounding networks. For direct communication of two people (i.e., their PANs), the bridging

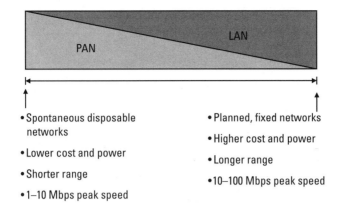

Figure 8.1 Complementary position of WLANs and WPANs (*From:* [5]).

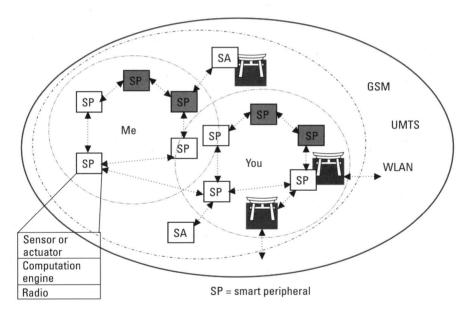

Figure 8.2 PAN network model.

functionality should be incorporated into each PAN. For communication through the external network, a PAN should implement routing and/or gateway functionalities. The standardization of topology and architecture are still open issues within the PAN at the time of writing. A layer-oriented scalable architecture should support functionalities and protocols of the first three layers and should provide the capability to communicate with the external world through higher layer connectivity. The middleware structure should be able to manage the system according to the access to the networks, resource discovery, support for scalability and reconfigurability, and QoS provisioning. It should also support downloadable applications. From the user point of view, PAN should offer plug-and-play connectivity.

8.2.2   Applications Scenarios and Service Requirements

Many different operating scenarios can be foreseen, mainly concentrated around:

- Personal services;
- Business services;
- Entertainment services.

Personal services include medical telemonitoring, control applications, and smart homes. Examples include:

- Medical professionals can constantly monitor the sick person whenever he/she moves.
- A person may use a small portable device to send commands to domestic appliances, such as actuators that control heating, curtains, windows, either from home or away.
- A person enters his car and can listen to his corporate e-mail text read by a computer, dictate, and send replies. This could be realized, for instance, by linking up and temporarily extending the person's PAN containing a 3G-enabled PDA with on-board speakers, microphones, and voice-recognition and synthesis systems.

Some examples of the business scenario include:

- The possibility for a person at the office to access a printer to print information stored on his device;
- Measuring of the temperature/pressure in the truck;
- Tracking the movement of employees within the offices.

Entertainment scenarios could consider high-tech applications, such as high-speed video on the train, or a game with head-mounted devices (Figure 8.3). A person can be carrying a small screen that shows a video, and whenever the user wants to watch the video on a larger screen (e.g., a laptop computer), it should be possible to pass the contents to this device in a seamless way.

Figure 8.3 Example of an entertainment PAN application: game with head-mounted device.

Therefore, applications will range from very low bit rate sensor and control data (10 bps) up to very high bit rate video streaming and bandwidth demanding interactive games (10 Mbps). The demands of QoS, availability, reliability, and security will differ greatly depending on the person and the activity a person is engaged in. Furthermore, the concept of QoS will gain different flavor due to the emergence of unique applications for which the timeliness and quickness of the ad hoc connection establishment is critically important, while the actual rate that the application will use over the already established connection may be quite modest.

8.2.3 Possible Devices

Many important and practical PAN applications require very low cost, potentially throw-away devices, capable of operating over an extended time without changing batteries, and with limited functionalities. Some other devices should incorporate advanced networking and computational functionalities and support high data rates. Some devices should be wearable or attached to the person (i.e., sensors); others could be stationary or associated temporally to the POS such as environmental sensors and printers. However, the most capable PAN devices should incorporate multimode functionalities to enable access to multiple networks. For these terminals, scalabilities with respect to the data rates, power consumption, supporting interfaces, and complexity are crucial issues. The radio interfaces of these devices will be different. Some will be capable of short-range communication using, for example, Bluetooth or IrDA; others will have WLAN interfaces or cellular radio interfaces.

Since they are basically battery-driven devices, one of most critical common requirement for all classes of PAN devices is the power efficiency.

8.3   State of the Art

The concept of PAN emerged at MIT in 1995 [1] as a way to have personal devices attached to the human body, which can communicate by using the intra-body electrical currents. This idea was first accepted and further developed at IBM Research [6]. A variety of PAN solutions are offered by using different existing radio technologies such as:

- Oxygen project (MIT) [7];
- Pico-radio [8];
- IrDA [9];
- HomeRF [10];
- Bluetooth [11].

The first versions of wearable electronics reached the markets in 2000 [12]. These early examples [13, 14] consist of a few electronic devices (phone, MP3 player, headset, microphone, and controller) that are directly connected together within a jacket. Some integration between devices is achieved so that, for instance, the music mutes when the phone rings and control of both phone and MP3 player is possible with a separate, easily accessible, controller. Washable wires, integrated into the garment, connect the devices. In the future we might expect both conducting fabrics and wireless links being used to make interdevice connections.

The IEEE 802.15 Working Group has the task of developing standards for these short-range wireless networks [15]. The WG is following very closely the work done by existing technologies that have already gained market interest and industry participation, especially Bluetooth. The Bluetooth wireless technology is an industry-based specification for a small form-factor, low-cost radio solution providing links between mobile computers, mobile phones, and other portable handheld devices, and connectivity to the Internet. Bluetooth operates in the 2.4-GHz ISM band. It uses FSK and slow frequency hopping. It supports 721 Kbps and it was originally developed as a cable replacement (10-m range) with transmission power of few mW. The connection of devices is supported by a star network topology.

In many respects, Bluetooth represents the state of the art in PAN technology. Some of its ideas are incorporated in the ongoing IEEE 802.15 standards concerning PANs.

8.3.1 Standardization Process

In March 1999, the IEEE P802.15 Working Group for WPANs was formed to develop standards for personal short-range wireless networks [16]. The scope of the work is to define the physical layer and MAC specifications for wireless POS connectivity. Therefore, P802.15 only addresses the bottom half of the ISO data link (see Chapter 2), which decides how and when the radio should be used for communication. The other half of the link layer has been standardized as IEEE P802.2 and maintains logical associations between the upper layers of the communication systems. All PHY and MAC standards use the same LLC in P802 [16].

The 802.15 WG works closely with the Bluetooth Special Interest Group (SIG). To prevent emerging two noninteroperable standards, the Bluetooth SIG imposed conditions on IEEE to ensure 100% interoperability with Bluetooth 1.0 and to include testing interfaces decided in Bluetooth 1.0 and new WPAN standards.

The IEEE standard 802.15.1-2002 was conditionally approved as a new standard by the IEEE-SA Standard Board on March 21, 2002. Several task groups (TGs) have been formed to face some crucial issues such as:

- Compatibility with the Bluetooth v1.1;

- Coexistence between WPANs and other wireless devices, such as the IEEE 802.11 WLANs, in the unlicensed 2.4-GHz frequency band;

- Provision of low-power, low-cost, short-range solutions targeted to consumer digital imagining and multimedia applications;

- Provision of very low power operation and very low complexity for low data rate applications such as sensors, interactive toys, smart budges, remote control, and home automation.

8.4 Technical Challenges of Future PANs

Although Bluetooth is relatively low cost and low power, it is not the optimum solution for several PAN applications, such as telemonitoring and control, which require very low data rate information (few hundred bits per second). Therefore, alternative technologies must be considered and the ultra wideband (UWB) technology is one of the most promising [17].

Suitable antennas must be developed for PAN applications with the following requirements [18]:

- Small and low cost;
- Electrically efficient in case of UWB;
- Large bandwidth and possible operation over multiple bandwidths.

Multiple antenna systems tailored to PAN applications and communications techniques can improve coexistence with other systems and overall capacity, enhancing the interference mitigation capabilities and the robustness of connectivity between a user PAN and external networks.

Most of the technical challenges that arise when WPAN is considered have been already addressed in the more general framework of short-range technologies, such as WLAN and ad hoc networks. So far, research in ad hoc networking has not been bound to any particular technology. Routing, security, and power efficiency issues should be addressed taking into account the peculiarities of the PAN ad hoc networks. For example, the restricted radio coverage area of a WPAN network with respect to a wide area ad hoc network can provide benefits from the security point of view; moreover, the mobility of nodes and dynamics are different in the two mentioned ad hoc networks. Furthermore, in some PAN applications requirements different than the speed rate are important, such as the capability to discover quickly a service and establish the connection. Finally, the development of WPAN technology should not only offer a solution superior to the existing ones, but it should also offer a glue among different technologies that will have gained widespread use at the time this "superior" solution will appear.

The rest of this section will highlight the main technical issues that arise in the context of WPAN systems, after recalling the basic concept of the emerging UWB technology.

8.4.1 UWB for WPAN

According to the FCC, the term "UWB technology" defines any wireless transmission scheme that occupies a bandwidth of more than 25% of a center frequency or more than 1.5 GHz. UWB appeared in the 1980s [5, 19–21] and it was mainly used for radar-based applications that could exploit accurate location and tracking capabilities. Due to the recent developments in low-cost, low-power switching technology and processing, UWB is becoming more attractive for low-cost communication devices. UWB is typically implemented in a carrierless fashion by directly modulating an "impulse" that has a very sharp rise and fall time.

Interesting features of UWB for PAN applications are as follows:

- *Great spatial capacity:* One UWB technology developer has measured peak speeds of 50 Mbps at a range of 10m and foresees that six of such systems could operate within the same 10-m radius circle with only minimal degradation [19]. This results in a spatial capacity of 1,000,000 bps/m^2, which is about 12 times that of IEEE 802.11a and 30 times that of a Bluetooth system [5].

- *Potential compliance for global unlicensed operation:* Power emissions of the UWB are targeted to be below the emission levels currently allowed for unintentional emitters. Therefore, UWB will coexist with and over-lay (or more precisely, underlie) existing narrow radio services while causing nearly imperceptible changes in the noise floor of those receivers.

- *High resolution and robust distance measuring capability:* The extremely fine time resolution of UWB systems enables the development of pre-cise localization capability. This capability can be used to provide new functionality for security and more generally a more efficient use of the network.

Furthermore, as compared with traditional radio transceiver architectures, the UWB transceivers are simpler and have low power consumption.

8.4.2 Power Efficiency

In the PAN environment the communication nodes are small and rely on limited battery energy for their operation. Much research has been performed aiming to reduce the energy consumption at the hardware level [22] and at different levels of the protocol stack [23], including work on routing [24–34], MAC [35–38], and transport protocols [39, 40].

A survey on some of the proposed solutions is presented in the rest of the section.

Power-Aware Routing

Power-aware routing algorithms select the route according to some predefined power cost functions [24–33]. In [28], the so-called minimum transmission energy (MTE) routing scheme is presented, which selects the route using the least amount of energy to transport a packet from source to destination. In [31], the function that is maximized is the network lifetime, which is defined as the period from the time instant when the network starts functioning to the time instant when the first node runs out of energy. In [34], the problem of finding the most beneficial source rate allocation and flow control strategy, given a required network lifetime, as defined in [31], is posed. Each source is associated with a utility function that increases with the traffic flowing over the available source-destination routes. The objective of the optimization problem is maximizing the sum of the source utilities for a required network lifetime guarantee. However, the definition of network lifetime used in [31] can be unsatisfactory, since it does not address some issues of ad hoc networks. For instance, the death of a single node does not completely disrupt the communication between the various source-destination pairs.

The definition of more appropriate performance metrics and algorithms design according to them is still an open research area [34].

Transmission Power Control

Transmission power control (TPC) has been extensively studied in the context of cellular networks, both TDMA/FDMA and CDMA-based [41–48]. The main motivation of these schemes has not been to conserve energy but to mitigate the effect of interference that one user may cause to others. Furthermore, most of the solutions rely on a centralized control and duplex communications that are not inherently present in wireless ad hoc networks. More recent works show, through theoretical studies [49] and simulations, that by applying TPC in ad hoc packet networks, considerable benefits in terms of capacity and energy consumption [50, 51] in a single-hop environment [52] and in a multihop environment [53] can be obtained. In [54] the TCP is employed to control the topology of wireless ad hoc networks. In [55, 56], power control is proposed as a part of the multiple access protocol for the class of CSMA/CA protocols and, more in general, for contention-based multiple access protocols. In the IEEE 802.11 standard for WLAN, the MAC technique is called distributed coordination function (DCF). DCF is a CSMA/CA scheme where retransmission of collided packets is managed according to binary exponential back-off rules (see Chapter 2). The standard also defines an optional point coordination function (PCF), which is a centralized MAC protocol able to support collision-free and time-bounded services. Power control is undesirable for DCF as the number of hidden terminals is likely to increase, which, in turn, results in more collisions and in more energy consumption. On the other hand, it can be effective in the PCF access mechanism, since there is no hidden terminal problem [50].

MAC Protocols

MAC protocols have the fundamental task of avoiding collisions, so that two interfering nodes do not transmit at the same time. Collisions are one cause of energy waste but not the only one. Other sources of energy waste are:

- *Control packet overhead:* Sending and receiving packets consumes energy;

- *Overhearing:* A node picks up packets that are destined to other nodes;

- *Idle listening:* Listening to receive possible packets that do not arrive.

Many measurements made for the IEEE 802.11 have shown that idle listening consumes 50 to 100% of the energy required for receiving [57]. A power-efficient MAC protocol should reduce the waste of energy from all the mentioned sources.

Various energy-efficient MAC have been proposed [35–38]. Some of them have been recalled in Chapter 2, where the power efficiency problem is presented mainly for cellular networks. In fact, most of the proposed solutions assume some kind of infrastructure network where there is a base station or an elected node that coordinates the action of the other nodes and helps them to control their power consumption. In a PAN environment, it is not always possible to assume that this node is available, and so other solutions have to be developed [23].

Furthermore, the need to support different service classes with some QoS guarantees typically requires some kind of scheduling. This is in contrast with the random nature of most current MAC protocols. Delicate trade-off is required between scheduled users and best-effort users. The development of random access protocols with some kind of service differentiation still poses many open issues.

The majority of power savings comes from the availability of hardware to power down selected system elements that are not required. An increasing number of devices can support low-power sleep modes. IEEE 802.11 has a power saving mode where a node only needs to be awaked periodically. Bluetooth [58] provides three different low-power modes: sniff, hold, and park.

Another aspect of power management is to increase the lifetime of the battery of a mobile node by using energy-efficient battery management techniques like a proper battery discharge policy [59]. A battery consists of many electrochemical cells from which power needs to be drained when the node transmits a packet. When a cell is allowed to rest in between discharge periods, it is able to recover part of its charge, thanks to the diffusion mechanisms. A battery discharge policy decides which cells (electrochemical cells) should serve the packet and which cells are allowed to rest. Analytical and simulation results have shown that discharge policies have significant impact on the battery lifetime [60, 61].

8.4.3 Service Discovery/Selection

An important feature of WPAN devices is the capability of discovering/selecting a service/resource that is available within its communication range. For example, a PDA should be able to find a printer within its proximity and, provided that some security conditions are satisfied, it should be able to use it as if the printer has been installed in the PDA software. In some applications, a quick service discovery and establishment of the connection are more important than the peak speed of the connection itself. Therefore, the issue of the service discovery is an important issue that so far has not received enough interest. The most well-known current service discovery protocols are Jini, the IETF Service Location Protocol (SLP), Microsoft's Universal Plug and Play (UPnP), and the Information Access Service (IAS) protocol defined for short-range infrared communications. An optimized Service Discovery Protocol (SDP) for ad hoc networks and resource-constrained devices is the Bluetooth SDP [62]. In [62], the use of semantic information associated with services has been proposed to enhance the SDP matching mechanism. WPAN service discovery mechanisms should take into account the low processing power and storage capabilities of the wearable devices. Energy consumption could be improved by reducing the amount of data transmitted and received in service discovery and registration requests, while maintaining all semantic information completely.

8.4.4 Security

Security should cover air interfaces, software operations and operating systems, and user profiles. Different techniques, such as encryption and clippering, TTP mechanisms, and agent technology, will be implemented. Security communications between the foreign PANs should be realized through the gatekeeping functionalities. The PAN security should offer:

- Full identity;
- Full anonymity;
- Data security;
- Integrity.

Special attention should be paid to security issues in PANs for several reasons. In fact, PANs are person-centered and may directly or indirectly reveal information about the owner of the PAN. The use of wireless technology and the ad hoc nature of the network make PANs vulnerable to eavesdropping and intrusion, with potentially many "foreign" devices getting connected into the PAN. The issue of charging (accounting) for particular services may also need to be investigated in the PAN context.

8.4.5 Ad Hoc Networking

PANs have an ad hoc character. New devices are added or are removed as the "owner" of the PAN moves around to different locations, different activities, and different environments. Therefore, a person should be able to enter the company space and to relay on WLAN, or to enter the car and communicate with the car through on-board devices, such as displays and voice input modules, without disrupting the ongoing connections. Moreover, when two people owning PANs meet, their networks may temporarily fuse to exchange information or join in a common activity. The research on PANs is, hence, strictly related to the research on ad hoc networks for which there is a lot of literature. Ad hoc networks have received substantial interest in the recent years [63, 64]. Several distinctive features make the task of ad hoc network design highly challenging:

- Nodes appearing and disappearing;

- Nodes moving around (mobility);

- Nodes changing their state (e.g., sleeping, awake, active, or temporarily unavailable for privacy reasons);

- Support of multihop communications (i.e., the links between some nodes must be established by relaying via a third node or a group of nodes);

- The wireless units are mainly battery-driven;

- Security problems are augmented due to the absence of a central network security administration [65, 66].

The need for new routing approaches becomes immediately obvious in ad hoc networking. Much work has been done in the formulation of distributed routing algorithms (see Chapter 3) and power efficient routing algorithms, as highlighted in Section 8.4.2.

8.4.6 Coexistence and Interference-Reduction Techniques

WLANs and WPANs are currently using the same frequency band (namely the 2.4-GHz ISM band), thus creating mutual inband colored noise [67]. This leads to interference between the two technologies and, hence, to the need of coexistence mechanisms. It is worth mentioning that neither Bluetooth nor WLAN was designed initially with specific mechanisms to combat the interference each creates to the other. When both technologies are operating at the same time but are separated by more than 3m, they do not normally interfere mutually at a high degree. However, within 3m, and especially within one-half meter, they can significantly degrade each other performance.

Coexistence mechanisms are under development in the Task Group 2 of IEEE 802.15 WG. Two classes of coexistence mechanisms have been defined:

collaborative and noncollaborative techniques. With collaborative techniques it is possible for the WPAN and the WLAN to exchange information to reduce the mutual interference. Collaborative techniques can be implemented when interfering devices are colocated in the same terminal. With noncollaborative techniques there is no way to exchange information between the two network systems and they operate independently.

An example of a collaborative coexistence mechanism is the so-called MAC enhanced temporal algorithm (MEHTA) scheme presented to the IEEE 802.15 WG [68]. This is the scheduling algorithm that has been proposed to mitigate the interference between an 802.15 device and an 802.11b device being colocated in the same terminal. MEHTA involves the use of a centralized controller, which monitors the 802.15 and the 802.11 traffic and allows exchange of information between the two radio systems. The controller works at the MAC layer and allows precise timing of packet traffic, thus avoiding interference between the two devices. 802.15 voice traffic has priority over WLAN packets, otherwise WLAN traffic is transmitted first. When there is voice traffic pending, WLAN packets are queued.

A noncollaborative algorithm is proposed in [69]. It does not require a centralized scheduler. The mechanism is based on a traffic-shaping technique. The proposed algorithm is performed at the WLAN stations in presence of an 802.15 voice link.

Some interference-reduction techniques to improve the performance of colocated Bluetooth and IEEE 802.11b are as follows:

- *Adaptive frequency hopping:* Frequencies of the pseudo random hopping sequence are dynamically changed. This method is used for Bluetooth devices. If a Bluetooth device knows on which channel a nearby 802.11 device operates, it can avoid hopping on that channel, thus avoiding interference. By using fewer hopping frequencies, however, Bluetooth performance turn out to be degraded. Another drawback of this approach is that it has to be approved by FCC regulations.

- *Dual-mode radio switching:* This is a time division approach—only one system is active at any given time. When 802.11b devices are transmitting, all Bluetooth devices are placed in power-save mode, and vice-versa. The actual implementation of this approach can be done in different ways.

- *Driver level switching and MAC level switching:* These are also time division approaches where the switching function is handled at the driver level or MAC level, respectively.

- *Transmit power control:* This was proposed for both Bluetooth [70] and 802.11b devices [71]. Reducing the average transmission power has a two-fold benefit: (1) increase of the capacity per area; and (2) reduction of the interference.

- *Adaptive fragmentation:* This can be used for 802.11 devices. In case of no interference, the largest packet size (1,500 bytes) should be used, since it provides the highest throughput. However, using smaller packets in situations with interference will reduce the probability of interference from a Bluetooth device, thus improving the performance.

These solutions are far from being completed. The problem of coexistence is still in the exploration and experimentation stage.

8.5 Broadband PAN

The broadband-PAN (B-PAN) is a future development of PAN towards the wideband adaptive novel techniques capable of broadband wireless communications. It will support applications up to 400 Mbps and will probably operate over 5-GHz or 60-GHz frequency bands [72]. B-PAN will use new technologies such as UWB, voice-over-B-PAN, and adaptive modulation in order to support adaptive QoS. Research on new radio interfaces for B-PAN is an open research area.

8.6 Conclusions

PAN is a key technology for future heterogeneous communication scenarios. Possible application scenarios for these short-range communications, together with the existing and emerging technologies to support them, have been presented. Low power consumption, operation in the unlicensed spectrum, coexistence between the WPAN and the WLAN, low cost, and small package size are some of the most important technical challenges to be faced.

The great volume of research work in ad hoc routing protocols should be put in the context of concrete MAC layer realization. The future work on the short-range wireless network must apply the routing protocols in a way that is adapted both to the channel access/transmission conditions and the application requirements.

The research in MAC layer development for PAN should focus on cross-layer optimization, ensuring that the MAC layer mechanisms include functionality to ensure guaranteed service levels (QoS) and power efficiency. This requires MAC layer awareness of the requirements put on the data streams originating from the network layer, as well as mechanisms for adjusting the parameters of the physical link. This means that the MAC layer should be able to exchange information with both the network layer and the physical layer.

Moreover, some kind of scheduling is required to support different service classes with each their own characteristics. To avoid scheduled users from overriding the best-effort users, a delicate trade-off is required. There are still many open issues related to the development of random access protocols able to differentiate according to user class.

References

[1]   Zimmerman, T. G., "Personal Area Networks (PAN): Near-Field Intra-Body Communication," M.S. thesis, MIT Media Laboratory, Cambridge, MA, 1995.

[2]   Prasad, R., and L. Gavrilovska, "Personal Area Networks," Keynote speech, *Proc. EUROCON*, Bratislava, Slovakia, July 2001, Vol. 1, pp. iii–viii.

[3]   Prasad, R. "Basic Concept of Personal Area Networks," WWRF, Kick-off Meeting, Munich, Germany, March 2001.

[4]   Niemegeers, I. G., R. Prasad, and C. Bryce, "Personal Area Networks," WWRF Second Meeting, Helsinki, Finland, May 10–11, 2001.

[5]   Leeper, D. G., "A Long-Term View of Short-Range Wireless," *IEEE Computer*, Vol. 34, No. 6, June 2001.

[6]   IBM PAN, http://www.almaden.ibm.com/cs/user/pan/pan.html.

[7]   MIT Oxygen project, http://oxygen.lcs.mit.edu/.

[8]   PicoRadio, http://www.gigascale.org.picoradio/.

[9]   Infrared Data Association (IrDA), http://www.irda.org.

[10]   HomeRF Working Group, http://www.homerf.org.

[11]   Bluetooth, http://www.bluetooth.com/.

[12]   Van Dam, K., S. Pitchers, and M. Barnard, "From PAN to BAN: Why Body Area Networks," WWRF Second Meeting, Helsinki, Finland, May 10–11, 2001.

[13]   http://www.research.philips.com/password/pw3/pw3_4.html.

[14]   http://www.levis-icd.com/.

[15]   IEEE 802.15, http://grouper.ieee.org/groups/802/15.

[16]   Siep, T. M., et al., "Paving the Way for Personal Area Network Standards: An Overview of the IEEE P802.15 Working Group for Wireless Personal Area Networks," *IEEE Personal Comm.*, Feb. 2000, pp. 37–43.

[17]   Siwiak, K., and L. Huckabee, "UWB Radio Physical Layer for WPAN Standards," WWRF Fifth Meeting, London, March 2002.

[18]   Hutter, A., Qin Xu, and J. Farserotu, "Antenna Technology for Nomadic Personal Area Networks," WWRF Fifth Meeting, London, March 2002.

[19]   Ultra-Wideband Working Group, http://www.uwb.org (current May 30, 2002).

[20] Rofheart, M., "XtremeSpectrum Multimedia WPAN PHY," IEEE 802.15.3 Working Group Submission, LaJolla, CA, July 2000, http://grouper.ieee.org/groups/802/15/pub/Download.html (current May 30, 2002).

[21] Foerster, J., et al., "Ultra-Wideband Technology for Short- or Medium-Range Wireless Communications," *Intel Technology Journal*, Q2, 2001, http://intel.com/technology/itj/q22001/articles/art_4.htm (current May 30, 2002).

[22] Simunic, T., et al., "Energy Efficient Design of Portable Wireless Systems," *Proceedings of the 2000 International Symposium on Low Power Electronics and Design*, Rapallo, Italy, July 25–27, 2000.

[23] Girling, G., et al., "The Design and Implementation of a Low Power Ad Hoc Protocol Stack," *Proceedings of IEEE Wireless Communications and Networking Conference*, Chicago, Sept. 2000.

[24] Xu, Y., J. Heidemann, and D. Estrin, "Geography-Informed Energy Conservation for Ad Hoc Routing," in *MOBICOM 2001*, July 2001.

[25] Chang, J.-H., and L. Tassiulas, "Energy Conserving Routing in Wireless Ad Hoc Network," in *INFOCOM 2000*, March 2000.

[26] Gomez, J., et al., "PARO: Power-Aware Routing in Wireless Packet Networks," in *Proc. 6th IEEE International Workshop on Mobile Multimedia Communications (MoMuC99)*, Nov. 1999.

[27] Feeney, L. M., "An Energy-Consumption Model for Performance Analysis of Routing Protocols for Mobile Ad Hoc Networks," *Baltzer/ACM Journal of Mobile Networks and Applications (MONET), Special Issue on QoS in Heterogeneous Wireless Networks*, June 2001.

[28] Shepard, T., "Decentralized Channel Management in Scalable Multihop Spread Spectrum Packet Radio Networks," Tech. Rep. MIT/LCS/TR-670, Massachusetts Institute of Technology Laboratory for Computer Science, July 1995.

[29] Singh, S., M. Woo, and C. S. Raghavendra, "Power-Aware Routing in Mobile and Ad Hoc Networks," *Proc. of 4th Annual ACM/IEEE International Conference on Mobile Computing and Networking*, Dallas, TX, Oct. 1998, pp. 181–190.

[30] Rodoplu, V., and T. H. Meng, "Minimum Energy Mobile Wireless Networks," *IEEE Journal on Selected Areas in Communications*, Vol. 17, No. 8, Aug. 1999, pp. 1333–1344.

[31] Chang, J.-H., and L. Tassiulas, "Energy Conserving Routing in Wireless ad hoc Networks," *Proc. of INFOCOM 2000*, Tel Aviv, Israel, March 2000.

[32] Manzoni, P., and J.-C. Cano, "A Performance Comparison of Energy Consumption for Mobile Ad Hoc Network Routing Protocols," *Proc. of the 8th International Symposium on Modeling, Analysis and Simulation of Computer and Telecommunication Systems, 2000*, pp. 57–64.

[33] Michail, A., and A. Ephremides, "Energy Efficient Routing for Connection-oriented Traffic in Ad Hoc Wireless Networks," *Proc. of the 11th IEEE International Symposium on Personal, Indoor and Mobile Radio Communications (PIMRC), 2000*, pp. 762–766.

[34] Srinivasan, V., et al., "Optimal Rate Allocation and Traffic Splits for Energy Efficient Routing in Ad Hoc Networks," in *INFOCOM 2000*, March 2002.

[35] Chockalingam, A., and M. Zorzi, "Energy Efficiency of Media AccessProtocols for Mobile Data Networks," *IEEE Transactions on Communications*, Vol. 46, No. 11, Nov. 1998.

[36] Sivalingam, K. M., M. B. Srivastava, and P. Agrawal, "Low Power Link and Access Protocols for Wireless Multimedia Networks," in *VTC '97*, May 1997.

[37] Monks, J. P., V. Bharghavan, and W. W. Hwu, "A Power Controlled Multiple Access Protocol for Wireless Packet Networks," in *INFOCOM 2001*, April 2001.

[38] Feeney, L. M., and M. Nilsson, "Investigating the Energy Consumption of a Wireless Network Interface in an Ad Hoc Networking Environment," in *INFOCOM 2001*, April 2001.

[39] Agrawal, S., and S. Singh, "An Experimental Study of TCP's Energy Consumption over a Wireless Link," in *4th European Personal Mobile Communications Conference*, Feb. 2001.

[40] Zorzi, M., and R. R. Rao, "Is TCP Energy Efficient?" in *Proc. IEEE MoMuC*, Nov. 1999.

[41] Bambos, N., "Toward Power-Sensitive Network Architectures in Wireless Communications: Concepts, Issues, and Design Aspects," *IEEE Personal Communications Magazine*, June 1998, pp. 50–59.

[42] Zander, J., "Distributed Cochannel Interference Control in Cellular Radio Systems," *IEEE Transactions on Vehicular Technology*, Vol. 41, No. 3, Aug.1992, pp. 305–311.

[43] Grandhi, S., et al., "Centralized Power Control in Cellular Radio Systems," *IEEE Transactions on Vehicular Technology*, Vol. 42, No. 4, Nov. 1993, pp. 466–468.

[44] Foschini, G., and Z. Miljanic "A Simple Distributed Autonomous Power Control Algorithm and Its Convergence," *IEEE Transactions on Vehicular Technology*, Vol. 42, No. 4, Nov. 1993, pp. 641–646.

[45] Grandhi, S., J. Zander, and R.Yates "Constrained Power Control," *International Journal of Wireless Personal Communications*, Vol. 1, No. 4, April 1995.

[46] Bambos, N., S. Chen, and G. Pottie "Radio Link Admission Algorithms for Wireless Networks with Power Control and Active Link Quality Protection," *Proc. IEEE INFOCOM '95*, 1995, pp. 97–104.

[47] Yates, R., "A Framework for Uplink Power Control in Cellular Radio Systems," *IEEE Journal on Selected Areas in Communications*, Vol. 13, No. 7, Sept. 1995, pp. 1341–1348.

[48] Ulukus, S., and R. Yates "Stochastic Power Control for Cellular Radio Systems," *IEEE Transactions on Communications*, Vol. 46, No. 6, June 1998, pp. 784–798.

[49] Gupta, P., and P. Kumar, "The Capacity of Wireless Networks," *IEEE Transactions on Information Theory*, Vol. 46, No. 2, March 2000, pp. 388–404.

[50] Qiao, D., et al., "Energy-Efficient PCF Operation of IEEE 802.11a Wireless LAN," *Proc. IEEE INFOCOM '02*, March 2002.

[51] Monks, J., V. Bharghavan, and W. Hwu, "Transmission Power Control for Multiple Access Wireless Networks," *Proc. IEEE Conference on Local Computer Networks LCN*, Nov. 2000, pp. 12–21.

[52]   Wu, S.-L., Y.-C. Tseng, and J.-P. Sheu, "Intelligent Medium Access for Mobile Ad Hoc Networks with Busy Tones and Power Control," *IEEE Journal on Selected Areas in Communications*, Vol. 18, No. 9, Sept. 2000, pp. 1647–1657.

[53]   Monks, J., et al., "A Study of the Energy Saving and Capacity Improvement Potential of Power Control in Multihop Wireless Networks," *Proc. IEEE Conference on Local Computer Networks LCN*, Nov. 2001, pp. 550–559

[54]   Ramanathan, R., and R. Rosales-Hain. "Topology Control of Multihop Wireless Networks Using Transmit Power Adjustment," *Proc. IEEE INFOCOM '00*, 2000.

[55]   Monks, J., V. Bharghavan, and W. Hwu, "A Power Controlled Multiple Access Protocol for Wireless Packet Networks," *Proc. IEEE INFOCOM '01*, April 2001.

[56]   ElBatt, T., and A. Ephremides, "Joint Scheduling and Power Control for Wireless Ad Hoc Networks," *Proc. IEEE INFOCOM '02*, 2002.

[57]   Ye, W., J. Heidemann, and D. Estrin, "An Energy-Efficient MAC Protocol for Wireless Sensor Networks," *Proc. IEEE INFOCOM '02*, 2002.

[58]   Woesner, H., et al., "Power-Saving Mechanisms in Emerging Standards for Wireless LANs: The MAC Level Perspective," *IEEE Personal Communications*, June 1998, pp. 40–48.

[59]   Adamou, M., and S. Sarkar, "A Framework for Optimal Battery Management for Wireless Nodes," *Proc. IEEE INFOCOM '02*, 2002.

[60]   Chiasserini, C. F., and R. R. Rao, "A Traffic Control Scheme to Optimize the Battery Pulsed Discharge," *Proc. of Milcom '99*, Atlantic City, NJ, Nov. 1999.

[61]   Chiasserini, C. F., and R. R. Rao, "Energy Efficient Battery Management," *Proc. of INFOCOM 2000*, Tel Aviv, Israel, March 2000.

[62]   Avancha, S., A. Joshi, and T. Finin, "Enhanced Service Discovery in Bluetooth," *Computer*, Vol. 35, No. 6, June 2002, pp. 96–99.

[63]   Toh, C.-K., *Wireless ATM and Ad Hoc Networks—Protocols and Architectures,* Boston, MA: Kluwer Academic Publishers, 1997.

[64]   Perkins, C., *Ad Hoc Networking*, Reading, MA: Addison Wesley, 2001.

[65]   Stajano, F., and R. Anderson, "The Resurrecting Duckling: Security Issues for Ad hoc Wireless Networks," in B. Christianson, B. Crispo, and M. Roe (eds.), *Security Protocols, 7th International Workshop Proceedings, Lecture Notes in Computer Science*, 1999.

[66]   Zhou, L., and Z. J. Haas, "Securing Ad Hoc Networks," *IEEE Network Magazine*, Vol. 13, No.6, Nov./Dec. 1999, pp. 24–30.

[67]   Lansford, J., A. Stephens, and R. Nevo, "Wi-Fi (802.11b) and Bluetooth: Enabling Coexistence," *IEEE Network*, Vol. 15, Issue 5, Sept.–Oct. 2001, pp. 20 –27.

[68]   Lansford, J., "MEHTA: A Method for Coexistence Between Co-Located 802.11b and Bluetooth Systems," IEEE 802.15-00/360r0, Nov. 2000.

[69]   Chiasserini, C. F., and R. R. Rao, "A Comparison Between Collaborative and Non-Collaborative Coexistence Mechanisms for Interference Mitigation in ISM Bands," *VTC 2001 Spring. IEEE VTS 53rd*, Vol. 3, 2001, pp. 2187–2191.

[70]   Eliezer, O., "Non-Collaborative Mechanisms for the Enhancement of Coexistence Performance," IEEE 802.15-01/092, Jan. 2001.

[71]   Shoemake, M. B., "Proposal for Power Control for Enhanced Coexistence," IEEE 802.15-01/081, Jan. 2001.

[72]   Karaoguz, J., "High-Rate Wireless Personal Area Networks," *IEEE Comm. Magazine*, Dec. 2001, pp. 96–102.

9

Future Vision

9.1 Introduction

The future of communication systems appears to be quite unpredictable. General requirements for future communication systems will mainly be derived from the types of service a user will require in the future. However, the future service and application scenario is not well defined. Services that will make UMTS a success are still unknown. Nevertheless, some key elements of future communication systems may be predicted and research trends defined [1–3].

9.1.1 User-Centric Scenario

Interesting visions for future communication scenarios can be found in [4–7]. In all of them, the new technologies are centered on the user, as it well shown by the reference multisphere user-centric model defined in "The Book of Visions— Visions of the Wireless World," [4] (see Figure 9.1).

At the first level, the user interacts with the elements closest to him/her or even part of the body. The second level of interaction involves elements belonging to the surrounding environment, such as computers and domestic appliances. This interaction should be easier, more friendly, and more personalized than it is today. As an example, the TV is expected to know our favorite programs. The third level of interaction is between the user and other people (e.g., people living in other countries) or entities, such as the car, through microphones, voice-recognition, and synthesis systems. Two fundamental steps to fulfill user requirements are the capability of accessing already available or future radio interfaces, as well as the possibility of interconnecting all different technologies. Finally, the user will also be able to get in touch with a self-created

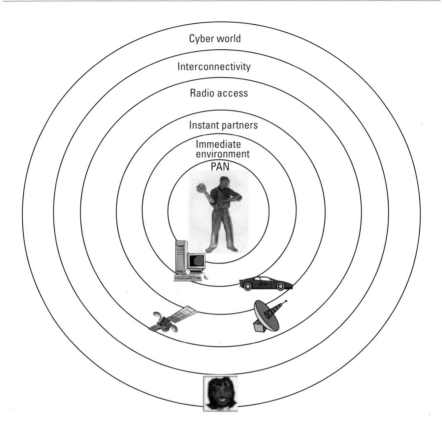

Figure 9.1 Reference multisphere model (*From:* [4]).

cyber world. Several video cameras or high-quality displays surrounding the user in the office or at home will allow him or her to start up a telepresence session via, for example, his or her PDA, where the user can have a virtual meeting with other real or not real people.

PAN technology (see Chapter 8) is a key technology of this user-centric scenario. In addition, two other concepts are emerging: (1) the cyber world, where the user can stay in touch with his agents, knowledge bases, communities, services, and transactions; and (2) the body area networks (BAN) concept, which goes beyond the PAN concept and consists of pervasive computing targeting environments where networked computing devices are ubiquitous and even integrated with the human user [8–10]. BAN devices will likely have to be able to use multiple physical links—for example, a mix of wired, RF, and infrared links. It should be possible to easily split a BAN device (e.g., the RF part from the keyboard and display) in order to attach only part of it to the body. This personal space expanded by virtual personal environments is the scope of the new concept of personal networks that goes beyond the PAN concept.

To make this vision possible, an *adaptive global network* has to be developed, which should provide flexible, adaptable, and secure, end-to-end communication and information services for:

- Small connections (to home and businesses);
- Smaller connections within homes, PANs, and between microrobots;

In this chapter, a scientific approach towards the realization of this future vision is described, and some research areas of interest are highlighted. Techniques and technologies introduced in this book provide the scientific background.

The concept of the adaptive global network and the key elements to be investigated are highlighted in Section 9.2. In the rest of the chapter, an approach for the investigation of each issue is proposed and the research areas of interest are outlined. Conclusions are drawn in Section 9.8.

9.2 Scientific Approach

In order to develop an adaptive global network as described in Section 9.1.1, the following key elements have to be investigated (see Figure 9.2):

- Adaptive and scalable air interfaces;
- Reconfigurable ambient networks;
- Security (across all layers);
- Optical network technologies;
- User interfaces and context-aware technologies;
- Flexible platforms.

The rest of the chapter consists of a more detailed description of the research to be done for each of the highlighted issues.

9.3 Adaptive and Scalable Air Interfaces

In order to justify the need for a new air interface, target requirements should be set high enough to ensure that the system will be able to serve long into the future. Considering the emerging technologies and the types of services that have to be provided in the future communication system, a reasonable approach would be to aim at 100-Mbps full mobility wide area coverage and 1-Gbps low

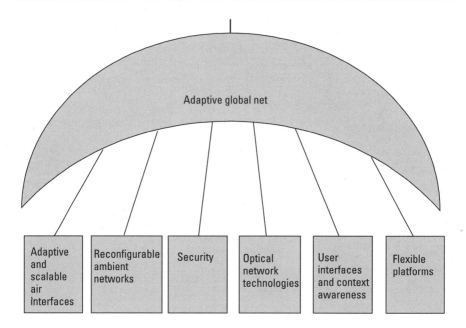

Figure 9.2   Research umbrella for achieving the objectives of the adaptive flexible future global network.

mobility local area coverage. Currently, the provision of highly reliable wireless link technologies envisages the use of OFDM or MC-CDMA, multiuser modulation, coding schemes, reconfigurable radio, smart antennas, power control, and advanced multiaccess technologies. To enable adaptivity to the network conditions and scalability (in terms of data rate and quality of service), new air interfaces (e.g., 3G+ like high speed downlink packet access, multicarrier and multitone CDMA, and ultra wide band) have to be investigated in terms of coexistence with other radio systems, multiple access capability, resistance to interferers, low cost, low power consumption, and implementation issues. The research-related areas are shown in Figure 9.3.

9.4   Reconfigurable Ambient Networks

To enable the advent of ambient intelligence, the underlying ambient networks need a high degree of reconfigurability. Reconfigurable networks imply the design of new communications mechanisms at different layers. These include research in the following areas (see Figure 9.4).

Intelligent Layered Resource Management Techniques

The reconfiguration actions have impact on various levels of mobile systems architecture and introduce high complexity that has to be handled by some

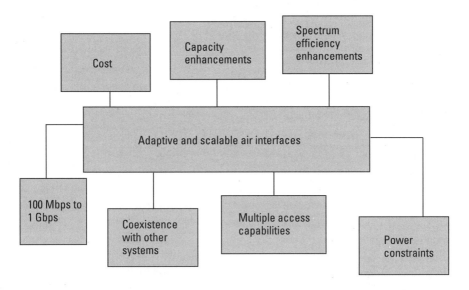

Figure 9.3 Research areas related to the development and validation of new air interfaces.

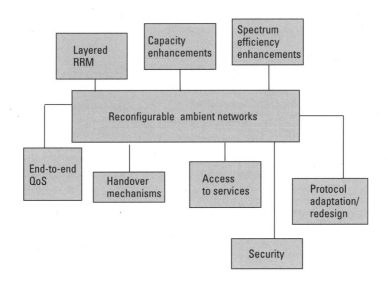

Figure 9.4 Research areas related to reconfigurable networks.

reconfigurability management intelligence (distributed or not). A possible approach is to start by describing evolutionary scenarios based on existing systems and gradually propose new scenarios that deploy leading-edge technologies. Research should start with the definition of user requirements, regulatory demands, system methodology, and constraints. Since reconfigurability concepts must be compatible to legacy networks and at the same time be embedded into future IP-based networks and future network topologies, such as ad hoc networks, the support by reconfigurable routers must be ensured.

Increase of Spectral Efficiency

Results derived from Radio Access Network (RAN) research and research in intelligent packet transfer techniques may be applied. This may help to identify new higher-layer (i.e., above the physical layer) concepts and algorithms to cope in an efficient manner with frequent topology variations, end user requirements, resource availability, and power constraint considerations. To this end, the development and optimization of the above layered and distributed resource management and radio link control algorithms, along with the appropriate middleware, can be conceived. Additionally, methods for spectrum sharing, mode detection, seamless service provision by mode switching, and secure software download in a combination of classical methods, such as mobile radio design and computer science, may be developed. This creates a synergy platform for novel concepts that can be combined with a full consideration of the real user needs.

Efficient Resource Management

A hierarchical management of all available resources can be achieved by a distributed monitoring along all the network segments, which evaluate system performance and detect bottlenecks. A resource management entity can apply resource management techniques for each system separately, increasing system performance. On top of the hierarchy, a management component can evaluate the load situation in all network segments and enable system interworking. This procedure obviously results in dynamic reconfiguration, since the management techniques that are applied in each network segment enable a real-time adaptation.

Protocol Adaptation

This might include in some cases the redesign of protocols, especially for multihop environments. To exploit the advanced physical layer and multiantenna solutions, including link adaptation and distributed packet scheduling, an internode radio resource management framework may be proposed. This is a first step towards a full cross-layer optimization (see also Chapter 5).

Furthermore, handover between diverse network technologies has to be investigated. This includes evaluation of the existing types of handover (i.e., soft, hard, interfrequency) and their interworking, also in case of a very large number of terminals; evaluation of handovers in different wireless systems; and evaluation of handovers between different wireless systems and provision for a seamless and secure handover mechanism by developing appropriate handover strategies.

Access to Services on Different Levels

Aspects such as dynamic software/protocol downloading are of importance for this item. The optimal provision of a service may necessitate installing dynamically specific software elements to the terminal or for some place in the network during service deployment or activation. For example, due to limited bandwidth available at the radio communications link, given service content (e.g., images, audio) should be drastically compressed so that it can be transmitted in real-time to the user terminal. For this purpose, an appropriate codec could be downloaded to a node at the edge of the mobile operator network as well as the terminal.

Security Across All Layers

To enable an end-to-end security in adaptive global network, the system will be characterized in terms of:

- Availability, which ensures the survivability of network services despite denial of service attack;

- Authentication, which enables a node to ensure the identity of the peer node it is communicating with;

- Confidentiality, which ensures that certain information never be disclosed to unauthorized entities—without proper authentication, of course, there is no point even to talk about confidentiality;

- Integrity, which guarantees that a message is never corrupted while it is transferred;

- Authorization, which is the process of deciding if device X is allowed to have access to service Y. It must be mentioned that devices that form ad hoc networks are more exposed to failure and theft and require specific types of security measures;

- Privacy, which can prevent the information of the users from flowing to others;

- Flexibility, which allows different upper protocols or applications to enforce their own security policy.

Furthermore, novel cryptography algorithms suitable for wireless communication and low-energy implementation should be developed.

9.5 Optical Network Technologies

Wireless systems have to be interconnected using a long-haul backbone network that is IP-based in order to provide global coverage. The network concepts are moving from network-centered towards person-centered solutions, introducing the new network paradigm of WPANs. The idea is illustrated in Figure 9.5.

The main concept is represented by the short-range network solution where the person within his or her personal space can establish connectivity to his or her personal devices and the outer world. The core network will be based on optical technology, complemented by a collection of other wired and wireless systems. In this sense, mechanisms for the provision of interoperability between existing and forthcoming wireless infrastructures with the optical one to support the IP flows become mandatory.

Research in the above frame includes:

- *Mobility mechanisms:* To ensure successful interaction among WPANs, the appropriate mechanism must be guaranteed for a dynamic type of communications.

- *Coexistence of WPANs and other wireless devices based on various standards:* This issue should be addressed by evaluating quantitatively the effects of coexistence in an initially established coexistence model consisting of the following four sections:

1. Physical layer models;
2. MAC layer models;
3. RF channel model;
4. Data traffic models.

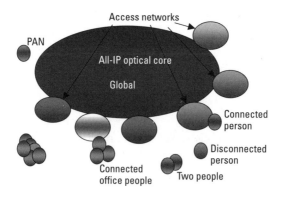

Figure 9.5 The future Internet.

An initial approach aims at understanding how different wireless services operating in the same band may affect each other. Later, coexistence mechanisms are established. Depending on the operating environment, one mechanism may be preferred over the other. The following information will serve as a determining factor whether a coexistence mechanism will have a positive impact:

- Complexity;
- Interoperability with systems not including a coexistence mechanism;
- Impact on interface to higher layers;
- Applicability to classes of operation;
- Voice and data support;
- Performance improvement of both systems;
- Impact on power management;
- Distributed resource control with QoS.

This will reflect the challenge of extending PAN by incorporating multi-mode terminals. The scenario is depicted in Figure 9.6.

Distributed resource control with QoS requires the following actions:

- Development of a common interface for exchanging QoS parameters;
- Resource monitoring in multistandard networks;

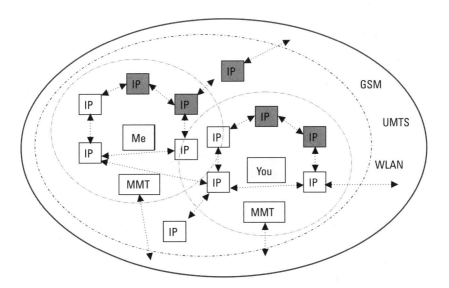

Figure 9.6 Distributed resource control with QoS.

- Distributed algorithms for efficient, timely and fair resource allocation;
- Developing new routing solutions for multistandard networks;
- Security.

A security system in such networks must be decentralized (i.e., it must support disconnected operations); flexible (i.e., it must support varying characteristics of hardware platforms); dynamic (i.e., it must be able to accommodate unpredictable network structure); and autonomous (i.e., it must be able to operate without network access). A functionality that is not found in conventional authorization systems but that will be useful for ad hoc networking environments is the possibility of specifying access conditions based on proximity, or on physical presence in a particular area. This implies a service providing secure location information.

9.6   Multimedia User Interfaces and Context-Aware Technologies

Interfaces like smart cards will play an important role in context-aware scenarios. The profile of the user (including not just the network parameters but also favorite URLs, e-mail address book, customer-bank profile, etc.) will be saved in the smart card through which the user will transparently interact with the communications infrastructures around him. In parallel, this interface will be used by operators as a billing support or for offering added value services, depending on location parameters, profile, etc.

Additionally, other ways of interaction, such as speech and/or multimodel interaction, have to be planned for these scenarios. Therefore, modules are needed that allow for the generation of spoken commands from textual input (synthetic speech) and enable recognition of spoken input (speech recognition), for the extraction of the meaning of the spoken input (spoken language understanding) and for maintaining an interactive dialog with the user. All these features are necessary for the creation of user-friendly spoken dialog-controlled communication between the application and the end user and finally for providing the user with the requested service (e.g., printing of a document using the local printer or receiving movie schedules for the local cinema). More research topics for speech, spoken language, and multimedia technologies exist in the context of deployment of handheld devices that will be used for short- or long-range communications. All services must be optimized to render the best-combined overall robustness and user-friendliness of the service at hand constrained by the variability of the network's end-to-end QoS, given that the network is ad hoc and adaptable.

Special attention is needed for the behavior of packet traffic in different network architectures, the reliability of the solutions, the control mechanisms and management of failure situations, and the configurability. More specifically, topics of study include different IP-over-photon solutions, utilization of optical switching techniques in IP infrastructure, and techniques that aim at an all-optical packet-switched network.

9.7 Flexible Platforms

In order to cope with the broad range of end user requirements in a context-aware environment, a flexible architecture has to been designed. Flexible architecture means that, to accommodate a wide spectrum of devices and functionalities in wireless systems, the underlying hardware platforms need a high degree of reconfigurability, including not only FPGA technology, but also mixed architectures based on a collection of microcontrollers, DSPs, FPGA, and hardware IP blocks. An appropriate design flow (including all algorithmic stages up to the system level) should enable the codesign of both various specific devices and flexible high-end multisystem devices. In order to ensure flexibility, taking into account various constraints (low power, low cost, low form factor), this design flow heavily relies on embedded systems and system-on-a-chip concepts. Last but not least, the introduction of multisystem devices is only possible with the help of a high degree of reconfigurable antennas and RF subsystems, and the help of research on new RF architectures (e.g., zero-IF). The different areas of relevance are shown in Figure 9.7.

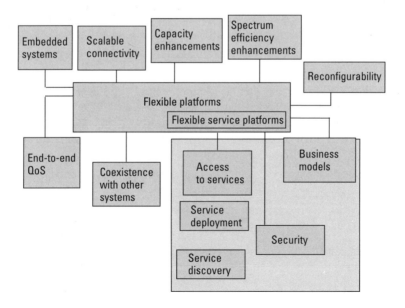

Figure 9.7 Development of flexible platforms.

An envisaged distributed antenna architecture, for example, will allow multioperator, multistandard radio transmission with investigation focusing on spectrum management and coexistence with other systems. The interworking between WLANs and other existing and future radio systems (e.g., GSM/UMTS, HIPERLAN/2, and UWB) will enable scalable wireless connectivity and can provide the stimulus to the development of future-generation radio interfaces. Economy, performance, interworking issues between different radio systems, and simplicity of the system in terms of operation and maintenance will be among the determining factors. Analysis and tests can be performed by appropriate system models and field trials to establish the most cost-effective implementations and to model the relationship between the optical design parameters and the systems performance.

9.8 Conclusions

A future vision on communications system cannot be complete without the definition of the future application scenario. A good starting point from which to draw trends for the future scenario is to look at our children who will be the active population in 10 years' time. The adventures of Tom Cruise in *Mission Impossible 1* and *2* are real for the kids that do take ubiquitous communication for granted. Another way to look at the future is through the science fiction stories of Arthur C. Clarke. In his paper "Extra Terrestrial Relays," published on the *Wireless World* in 1945 [11], he envisaged the possibility of using repeaters flying in space for wireless communications. With this revolutionary idea, he has become the father of satellite communications. We have to be ready for radical changes in the environment. In this context, the classical 10-year plan engineering approach of looking at the future should be abandoned to give room to an evolutionary/chaotic approach where the user stands at the center. While this book aims to provide the reader with a background on the existing and emerging technologies for future wireless communications systems, it is more and more evident that concurrency of the classical electronic, communication, math, and engineering disciplines, with other disciplines like business, sociology, and psychology is needed to draw a fully successful application scenario.

References

[1]   Huomo, H., E. Cianca, and R. Prasad, "A Lesson on Unpredictable Future," *Wireless Personal Communications,* Vol. 22, No. 2, August 2002, pp. 331–336.

[2]   *Proceedings Fourth Strategic Workshop*, Prague, Czech Republic, Sept. 6–7, 2002.

[3]   Lillenberg, J., and R. Prasad, "Research Challenges for 3G and Paving the Way for Emerging New Generations," *Wireless Personal Communication,* Vol. 17, June 2001, pp. 355–362.

[4] Mohr, W., (ed.), et al., "The Book of Visions 2000—Visions of the Wireless World," Version 1.0, *Wireless Strategic Initiative*, www.wireless-world-research.org, Nov. 2000.

[5] Zander, J., et al., "Telecom Scenario's in 2010," PCC, KTH, Sweden, 1999.

[6] ACM, "The Next 1000 Years," Special Issue of Communications of the ACM, Vol. 44, No. 3, March 2001.

[7] Niemegeers, I. G., and S. M. Heemstra De Groot, "From Personal Area Networks to Personal Networks: A User Oriented Approach," *Wireless Personal Communications*, Vol. 22, No. 2, August 2002, pp. 175–186.

[8] Gupta, S. K., et al., "An Overview of Pervasive Computing," Guest editorial, *IEEE Personal Communications*, Vol. 8, No. 4, Aug. 2001, pp. 8–9.

[9] Van Dam, K., S. Pitchers, and M. Barnard, "Body Area Networks—Towards a Wearable Future," WWRF Kick-off meeting, Munich, Germany, March 6–7, 2001.

[10] Hum, A. P. J., "Fabric Area Network—A New Wireless Communications Infrastructure to Enable Ubiquitous Networking and Sensing on Intelligent Clothing," Special issue on Pervasive computing and computer networks (IBM), 2000.

[11] Clarke, A. C., "Extra Terrestrial Relays," *Wireless World*, Oct. 1945, pp. 305–308.

List of Acronyms

1G	first generation
2G	second generation (2G)
3G	third generation (3G)
AAA	authentication, accounting, authorization
AAL5	ATM adaptation layer 5
ACK	acknowledgment
AGC	automatic gain control
ANSI	American National Standard Institute
AMPS	Advanced Mobile Phone Services
ARIB	Association of Radio Industries and Broadcasting
ARP	Address Resolution Protocol
ARQ	automatic repeat request
ATM	Asynchronous Transfer Mode
AWGN	additive white Gaussian noise
BAN	body area networks
BD	Bandwidth * Delay product
BER	bit error rate
B-PAN	broadband-PAN

BPS	bit per symbol
BRAN	Broadband Radio Access Network
BS	base station
BSC	binary symmetric channel
BTMA	busy tone multiple access
BTS	base transceiver station
CA	collision avoidance
CAC	call admission control
CC	congestion control
CD	collision detection
CDMA	code division multiple access
CDPD	cellular digital packet access
CE	controlled equalization
CIR	carrier-to-noise ratio
CN	core network
COA	care-of address
COFDM	coded-OFDM
CPC	complementary punctured convolutional codes
CPRMA	centralized-PRMA
CRC	cyclic redundancy check
CS	circuit switched
CSF	contrast sensitivity function
CSI	channel side information
CSMA	carrier sense multiple access
CWDN	congestion window
DAB	Digital Audio Broadcasting
DAMA	demand assigned multiple access
DCA	dynamic channel assignment
DCF	Distributed Coordination Function

DCH	dedicated channel
DCT	discrete cosine transform
DFE	decision feedback equalizer
DFT	discrete Fourier transform
DHA	dynamic home address
DiffServ	differentiated services
DL	downlink
DNS	domain name system
DPCM	differential pulse code modulation
DPDCH	dedicated physical data channel
DPRMA	dynamic packet reservation multiple access
DS-CDMA	direct sequence CDMA
DSMA	digital sense multiple access
DVB-T	Digital Video Broadcasting
DVD	Digital Versatile Disc
DVMRP	Distance Vector Multicast Routing Protocol
EDGE	Enhanced Data Rates for GSM Evolution
ELN	explicit loss notification
ESN	electronic serial number
ESP	encapsulating security payload
ETSI	European Telecommunications Standards Institute
FA	foreign agent
FCA	fixed channel assignment
FCC	Federal Communications Commission
FCH	forward access channel
FDD	frequency division duplex
FEC	forward error correction
FFT	fast Fourier transform
FGS	fine granularity scalability

FH	frequency hopping
FPLMTS	Future Public Land Mobile Telecommunications System
GBN	Go-back-N
GE	Gilbert-Elliott
GEO	geostationary Earth orbit
GGSN	gateway GPRS support node
GPRS	General Packet Radio Service
GPS	Global Positioning System
GRE	generic routing encapsulation
GSM	Global System for Mobile Communications
GW	gateway
HA	home agent
HARQ	hybrid ARQ
HAWAII	Handoff-Aware Wireless Access Internet Infrastructure
HDAA	Home Domain Allocation Agency
HIPERLAN	High Performance Local Area Network
HLR	home location register
HSCSD	High-Speed Circuit-Switched Data
HTML	hypertext markup language
HVS	human visual system
IAS	information access service
ICMP	Internet Control Message Protocol
IETF	Internet Engineering Task Force
IEEE	Institute of Electrical and Electronics Engineering
IGMP	Internet Group Management Protocol
IM	intermodulation
IMT-2000	International Mobile Telecommunications-2000
IntServ	integrated services
IP	Internet Protocol

ISDN	Integrated Services Digital Network
ISI	intersymbol interference
ISMA	inhibit sense multiple access
ISO/OSI	International Standard Organization/Open Systems Interconnect
IXC	interexchange carrier
JEIC	joint multiple access interference canceler/equalizer
JND	just-noticeable distortion
JPEG	Joint Photographic Experts Group
LAN	local area network
LEO	low-Earth orbit
LFN	long fat network
LLC	link layer control
LSP	label switched path
LSR	label-switching router
LTN	long thin networks
MAC	multiple access control
MAI	multiple access interference
MANET	Mobile Ad Hoc Networking
MAP	multiple access protocols
MBM	motion boundary marker
MC-CDMA	multicarrier CDMA
MCER	motion compensated error residual
MDC	multiple description coding
MD-PRMA	multidimensional PRMA
MIMO	multiple input–multiple output
MMAC	multimedia mobile access communications
MMSEC	minimum mean square error combining
MN	mobile node
MND	minimally noticeable distortion

MPEG	Moving Pictures Experts Group
MPLS	multi-protocol label switching
MRC	maximum ratio combining
MSC	mobile switching center
MT-CDMA	multitone CDMA
MTE	minimum transmission energy
MUD	multiuser detection
MV	motion vector
NACK	negative-acknowledgment
NAI	network access identifier
NO	network operators
OBI	observation interval
OFDM	orthogonal frequency division multiplexing
ORC	orthogonal restoring combining
OSI	Open Systems Interconnection
OVSF	orthogonal variable spreading factor
PAM	pulse amplitude modulation
PAN	personal area network
PCF	point coordination function
PDA	personal digital assistant
PDC	Personal Digital Cellular System
PDN	packet data network
PDU	packet data unit
PER	packet error rate
PFGS	progressive fine granularity scalability
PHB	per-hop behavior
PHY	physical layer
PIM	protocol independent multicast
PKI	public key infrastructure

PN	pseudo noise sequences
POS	personal operating space
PRMA	packet reservation multiple access
PSK	phase shift keying
PSNR	peak signal-to-noise ratio
PSTN	Public Switched Telephone Network
QAM	quadrature amplitude modulation
QoS	quality of service
RAB	radio access bearer
RACH	random access channel
RCC	routing control center
RCPC	rate-compatible convolutional codes
RF	radio frequency
RIP	Routing Information Protocol
RLC	radio link control
RNS	radio network subsystem
RRC	radio resource control
RRM	radio resource management
RSVP	Reservation Protocol
RTCP	Real-Time Control Protocol
RTO	retransmission time-out
RTP	Real-Time Transport Protocol
RTSP	Real-Time Streaming Protocol
RTT	round-trip time
Rx	receiver
SA	security association
SACK	selective acknowledgements
SDP	Service Discovery Protocol
SGSN	service GPRS support node

SIP	Session Initiation Protocol
SIR	signal-to-interference ratio
SLA	service level agreement
SLP	Service Location Protocol
SNR	signal-to-noise ratio
S/P	serial-to-parallel
SQ	scalar quantization
SR	selective repeat
SRE	source route entry
SSA	secure scaleable authentication
SSL	secure sockets layer
SSTHRESH	slow start threshold
STP	Satellite Transport Protocol
TBF	temporary block flow
TCM	trellis-coded modulation
TCP	Transport Control Protocol
TDD	time division duplex
TDMA	time division multiple access
TF	transmission formats
TFI	temporary flow identity
TOS	type of service
TTC	Telecommunications Technology Council
TTL	time-to-live
TTP	trusted third parties
Tx	transmitter
UDP	User Datagram Protocol
UEP	unequal error protection
UL	uplink
UMTS	Universal Mobile Telecommunications System

UTRA	UMTS Terrestrial Radio Access
UTRAN	UMTS Radio Access Network
UWB	ultra wideband
VLR	visitor location register
VLSI	very large scale integration
VoIP	voice over IP
VP/VC	virtual path/virtual channel identifiers
VQ	vector quantization
WAP	Wireless Application Protocol
WARC	World Administrative Radio Conference
WCDMA	wideband CDMA
WG	working group
WH	Walsh-Hadamard
WLAN	wireless LAN
WPAN	wireless personal area network
WWAN	wireless wide area network

About the Authors

Ramjee Prasad received his B.Sc. (eng.) from the Bihar Institute of Technology, Sindri, India, and his M.Sc. (eng.) and Ph.D. from Birla Institute of Technology (BIT), Ranchi, India, in 1968, 1970, and 1979, respectively.

Professor Prasad joined BIT as a senior research fellow in 1970 and became an associate professor in 1980. While he was with BIT, he supervised a number of research projects in the area of microwave and plasma engineering. From 1983 to 1988, he was with the University of Dar es Salaam (UDSM), Tanzania, where he became a professor of telecommunications in the Department of Electrical Engineering in 1986. At UDSM, he was responsible for the collaborative project Satellite Communications for Rural Zones with Eindhoven University of Technology, the Netherlands. From 1988 through 1999, he was with the Telecommunications and Traffic Control Systems Group at Delft University of Technology (DUT), where he was actively involved in the area of wireless personal and multimedia communications (WPMC). He was the founding head and program director of the Center for Wireless and Personal Communications (CEWPC) of International Research Center for Telecommunications—Transmission and Radar (IRCTR). Since 1999, Professor Prasad has been with Aalborg University, as the codirector of the Center for Person Kommunikation (CPK), and holds the chair of wireless information and multimedia communications. He was involved in the European ACTS project Future Radio Wideband Multiple Access Systems (FRAMES) as a DUT project leader. He is a project leader of several international, industrially funded projects. He has published more than 300 technical papers, contributed to several books, and has authored, coauthored, and edited 12 books: *CDMA for Wireless Personal Communications, Universal Wireless Personal Communications, Wideband CDMA for Third Generation Mobile Communications, OFDM for Wireless Multimedia Communications, Third Generation Mobile*

Communication Systems, WCDMA: Towards IP Mobility and Mobile Internet, Towards a Global 3G System: Advanced Mobile Communications in Europe, Volumes 1 & 2, IP/ATM Mobile Satellite Networks, Simulation and Software Radio for Mobile Communications, Wireless IP and Building the Mobile Internet, and *WLANs and WPANs towards 4G Wireless,* all published by Artech House. His current research interests lie in wireless networks, packet communications, multiple-access protocols, advanced radio techniques, and multimedia communications.

Professor Prasad has served as a member of the advisory and program committees of several IEEE international conferences. He has also presented keynote speeches and delivered papers and tutorials on WPMC at various universities, technical institutions, and IEEE conferences. He was also a member of the European cooperation in the scientific and technical research (COST-231) project dealing with the evolution of land mobile radio (including personal) communications as an expert for the Netherlands, and he was a member of the COST-259 project. He was the founder and chairman of the IEEE Vehicular Technology/Communications Society Joint Chapter, Benelux Section, and is now the honorary chairman. In addition, Professor Prasad is the founder of the IEEE Symposium on Communications and Vehicular Technology (SCVT) in the Benelux, and he was the symposium chairman of SCVT'93.

In addition, Professor Prasad is the coordinating editor and editor-in-chief of the *Kluwer International Journal on Wireless Personal Communications* and a member of the editorial board of other international journals, including the *IEEE Communications Magazine* and *IEE Electronics Communication Engineering Journal.* He was the technical program chairman of the PIMRC'94 International Symposium held in The Hague, the Netherlands, from September 19–23, 1994 and also of the Third Communication Theory Mini-Conference in Conjunction with GLOBECOM'94, held in San Francisco, California, from November 27–30, 1994. He was the conference chairman of the 50th IEEE Vehicular Technology Conference and the steering committee chairman of the second International Symposium WPMC, both held in Amsterdam, the Netherlands, from September 19–23, 1999. He was the general chairman of WPMC'01 which was held in Aalborg, Denmark, from September 9–12, 2001.

Professor Prasad is also the founding chairman of the European Center of Excellence in Telecommunications, known as HERMES. He is a fellow of IEE, a fellow of IETE, a senior member of IEEE, a member of the Netherlands Electronics and Radio Society (NERG), and a member of IDA (Engineering Society in Denmark).

Marina Ruggieri graduated cum laude in electronics engineering in 1984 from the University of Roma La Sapienza. She was with FACE-ITT in the high frequency division from 1985–1986, during which time she was trained on GaAs monolithic design and fabrication techniques at GTC-ITT (Roanoke, Virginia).

Professor Ruggieri was a research and teaching assistant at the University of Roma Tor Vergata in the electronics engineering department (1986–1991). She was an associate professor of signal theory at the University of L'Aquila (Department of Electrical Engineering, 1991–1994) and of digital signal processing at the University of Roma Tor Vergata (Department of Electronics Engineering, 1994–2000), also teaching radio communications systems (1994– 1999) and telecommunications systems (1999–2000) at the University of L'Aquila.

Since 2000 she has been a full professor of telecommunications at the University of Roma Tor Vergata (department of electronics engineering). Her teaching modules are digital signal processing, information and coding, and telecommunications signals and systems.

She has participated in international committees for the assignment of professor chair, Ph.D., and master's degrees in various universities (Lund-Sweden, Delft—the Netherlands, Toulouse-France, Trondheim-Norway, Aalborg-Denmark).

In 1999 she was appointed to the board of governors of the IEEE Aerospace and Electronics System Society (2000–2002) and reelected for 2003–2005.

Professor Ruggieri's research mainly concerns space communications systems (in particular, satellites) as well as mobile and multimedia networks.

She is the principal investigator of a satellite scientific communications mission [DAta and Video Interactive Distribution (DAVID)], of the Italian Space Agency, to be launched in 2004.

She is also the principal investigator of a 2-year national research program (CABIS) on CDMA integrated mobile systems, cofinanced by MIUR (2000-2002).

She has participated in various ESA contracts on the study and design of advanced configurations for satellite antennae at Ku-band, frequency scanning array for satellite communication, robust modulation, and coding for personal communication systems.

She is and has been involved in the organization of international conferences and workshops: the Third Workshop on Mobile/Personal Satcoms—EMPS 1998 (chair); the First Workshop on Strategic Research Plan for New Millenium Wireless World—SW 2000 (cochair); EMPS 2002 (vice chair); SW 2002 (cochair); IEEE Aerospace Conference 2002 and 2003 (track cochair); WPMC 2002, IEEE GLOBECOM 2002 Satellite Workshop; IEEE VTC Fall 2002 (TCP member) WPMC 2004 (TCP chair).

She is the editor of the *IEEE Transactions on AES* for "Space Systems." She is a member of the Editorial Board of *wireless personal communications*—an International Journal (Kluwer).

Professor Ruggieri was awarded the 1990 Piero Fanti International Prize and was nominated for the Harry M. Mimmo Award in 1996 and the Cristoforo Colombo Award in 2002.

She is an IEEE senior member (S'84-M'85-SM'94) and chair of the IEEE AES Space Systems Panel.

She is the author of approximately 150 papers on international journals, transactions, and proceedings of international conferences, book chapters, and books.

Index

UMTS and Mobile Computing, Alexander Joseph Huber and Josef Franz Huber

Understanding Cellular Radio, William Webb

Understanding Digital PCS: The TDMA Standard, Cameron Kelly Coursey

Understanding GPS: Principles and Applications, Elliott D. Kaplan, editor

Understanding WAP: Wireless Applications, Devices, and Services, Marcel van der Heijden and Marcus Taylor, editors

Universal Wireless Personal Communications, Ramjee Prasad

WCDMA: Towards IP Mobility and Mobile Internet, Tero Ojanperä and Ramjee Prasad, editors

Wireless Communications in Developing Countries: Cellular and Satellite Systems, Rachael E. Schwartz

Wireless Intelligent Networking, Gerry Christensen, Paul G. Florack, and Robert Duncan

Wireless LAN Standards and Applications, Asunción Santamaría and Francisco J. López-Hernández, editors

Wireless Technician's Handbook, Andrew Miceli

For further information on these and other Artech House titles, including previously considered out-of-print books now available through our In-Print-Forever® (IPF®) program, contact:

Artech House	Artech House
685 Canton Street	46 Gillingham Street
Norwood, MA 02062	London SW1V 1AH UK
Phone: 781-769-9750	Phone: +44 (0)20 7596-8750
Fax: 781-769-6334	Fax: +44 (0)20 7630-0166
e-mail: artech@artechhouse.com	e-mail: artech-uk@artechhouse.com

Find us on the World Wide Web at:
www.artechhouse.com